AF412892

DEVELOPMENTS IN POLYMERISATION—3

Network Formation and Cyclisation in Polymer Reactions

CONTENTS OF VOLUMES 1 AND 2

DEVELOPMENTS IN POLYMERISATION—3

Network Formation and Cyclisation
in Polymer Reactions

Edited by

R. N. HAWARD, Sc.D.

Emeritus Professor of Industrial Chemistry,
Department of Chemical Engineering, University of Birmingham, UK

APPLIED SCIENCE PUBLISHERS
LONDON and NEW JERSEY

APPLIED SCIENCE PUBLISHERS LTD
Ripple Road, Barking, Essex, England

APPLIED SCIENCE PUBLISHERS INC.
Englewood, New Jersey 07631, USA

British Library Cataloguing in Publication Data

Developments in polymerisation—3—(The Developments series)
1. Polymers and polymerisation—Periodicals
I. Series
547'.28'05   QD281.P6

ISBN 0-85334-117-6

WITH 19 TABLES AND 52 ILLUSTRATIONS

Printed in Great Britain by Galliard (Printers) Ltd, Great Yarmouth

PREFACE

The formation of polymer networks, in which essentially all the polymer molecules are connected together to form a continuous mesh, is a process of major importance in polymer science. Perhaps the most important commercial example of the process is the vulcanisation of rubber, but cross-linking, as it is called, also plays a vital part in the preparation of ion exchange resins, various types of polyester and polyurethane. In fact, as will be clear from the following chapters, a great number of different chemical reactions are employed for cross-linking purposes. They may, however, be conveniently divided into four groups:

(1) *step-wise addition reactions*, where the polymer molecules steadily grow in size along with the formation of cross-links (Chapter 3)

(2) *vinyl–divinyl polymerisations*, mainly by free-radical processes where high polymer is formed at once and may be subsequently joined together as the proportion of the monomers converted to high polymer increases (Chapter 4)

(3) *the cross-linking of preformed polymer*, as with rubber vulcanisation and radiation cross-linking (Chapters 2 and 5)

(4) *cyclopolymerisation*, where divinyl compounds are fully cyclised with negligible cross-linking (Chapter 1)

Clearly in group 4 no network is formed, but even in groups 1–3 there are many problems governing the extent to which the formation of chemical bonds actually leads to the creation of an extended network. Where this takes place during a polymerisation process there is generally an observable 'gel point' which occurs quite suddenly and where the reactants cease to flow reversibly as a liquid.

The theory of 'gel points' was first developed by Flory[1] and is summarised in his *Principles of Polymer Chemistry*. This work provides the starting point for most theories of network formation. Naturally there have been modifications of some of Flory's proposals in later work and it may be useful here to highlight the case of the radical copolymerisation of vinyl and polyvinyl compounds where a greater understanding of the reaction process has led to the most drastic amendments to the early proposals, and, indeed, provides an important justification for the present book.

Flory proposed that the gel point is reached when the expectancy of finding a cross-link in a particular existing molecule is unity. Considering, for example, the case of methacrylate–dimethacrylate copolymerisation, where an equal reactivity of acrylic groups could reasonably be assumed, he proposed that incipient gelation takes place when

$$\theta_c = \frac{1}{\rho_0 \bar{\gamma}_w} \tag{1}$$

where θ_c is the fraction of monomer molecules polymerised, ρ is the fraction of methacrylate groups occurring in dimethyacrylate units and $\bar{\gamma}_w$ is the weight average degree of polymerisation of the primary (zero conversion) polymer molecules.

Although this view is essentially correct in terms of the structure required to form a gel, and still provides the starting point for most cross-linking theories, it does not take into account the reaction steps by which such a structure is to be reached and especially it does not consider the possibility of intra-molecular cyclisation.

Over the years numerous efforts have been made to amend the classical cross-linking model to take account of cyclisation and in many cases, including stepwise polymerisations (Chapter 3) and rubber vulcanisation (Chapter 5), considerable progress has been made. For instance, the concept of 'wasted cross-links' has proved to be very useful.

The extreme example of this effect is that of cyclopolymerisation as described in Chapter 1. In these cases cyclisation is sufficiently dominant for a polymer prepared by polymerisation of a divinyl compound to be completely soluble and for no observable gel to be formed. This process was first reported by Butler and Angelo[2] and was certainly unexpected at the time, although Simpson and co-workers[3] had already reported extensive internal cyclisation during the polymerisation and gelation of diallyl esters. However, the occurrence of true cyclopolymerisation, which clearly requires a particularly stereochemical structure in the monomer, has now been extensively verified and studied as described in Chapter 1.

Chapters 2 and 5 describe the technically important processes of photo-initiated cross-linking and rubber vulcanisation where the starting point in each case is the long chain polymer molecule. Here, as already mentioned, cyclisation may be regarded as involving a certain restriction on cross-linking efficiency, but the main interest centres on the chemistry of the process rather than on the geometry of the cross-links.

It is when a polymer is built up *ab initio* from low molecular weight compounds that cyclisation and hence departure from the classical theory becomes most important. In stepwise addition polymerisations cyclisation generally occurs to a substantial extent but the actual extension of the polymer molecules in space still remains reasonably random as increasingly complex units join together. However, with free radical polymerisations the original conception of a statistical long chain with unreacted pendant double bonds is proving increasingly difficult to sustain. Indeed it has become clear, especially when polymerisation takes place in solution, that the very first polymer molecules to be formed (zero conversion polymer) are already cyclised to such an extent that their conformation is drastically modified and a much denser molecular coil is formed in the presence of a divinyl compound. Within such a coil, many, perhaps most, of the pendant vinyl groups are shielded from further reaction so that the joining together of different coils is impeded. Reaction then proceeds by the formation of new polymer molecules and by the attachment of polymeric radicals to existing coils, which may then develop into 'microgels'. These will ultimately be linked together and give a gel point. However, since the closeness of the polymer configuration in each microgel is determined by the extent of cyclisation and this increases with the proportion of cross-linking agent, a situation can arise where the gel point is independent of the initial concentration of cross-linker. This situation is clearly in conflict with the predictions of eqn. 1.

A fairly complete account of these effects is given in Chapter 4 which describes how our understanding of free radical network formation has developed rapidly in recent years. There are, however, still many un-answered questions which make the subject of rather greater interest than many polymer scientists, working in other fields, probably appreciate. It is hoped that the present book will encourage further research in the overall field of the cyclisation and cross-linking of polymers.

REFERENCES

1. FLORY, P. J., *Principles of Polymer Chemistry*, Ithaca, New York, Cornell Univ. Press, 1953, p. 391.
2. BUTLER, G. B. and ANGELO, R. J., *J. Amer. Chem. Soc.*, **79** (1957), p. 3129.
3. SIMPSON, W., HOLT, T. and ZEITE, R., *J. Polym. Sci.*, **10** (1953), p. 489.

CONTENTS

LIST OF CONTRIBUTORS

G. B. BUTLER

Professor, Department of Chemistry, and Director, Centre for Macro-molecular Science, University of Florida, 420 Bryant Space Sciences Building, Gainesville, Florida 32611, USA.

G. C. CORFIELD

Senior Lecturer, Department of Chemistry, Sheffield City Polytechnic, Pond Street, Sheffield S1 1WB, UK.

K. DUŠEK

Senior Research Scientist, Institute of Macromolecular Chemistry, Czechoslovakia Academy of Science, 16206 Prague 6, Czechoslovakia.

A. LEDWITH

Professor, Department of Inorganic, Physical and Industrial Chemistry, The University of Liverpool, Vine Street, Liverpool L69 3BX, UK.

M. MORTON

Regents Professor Emeritus of Polymer Chemistry, Institute of Polymer Science, University of Akron, Akron, Ohio 44326, USA.

R. F. T. STEPTO

Reader in Polymer and Fibre Science, Department of Polymer and Fibre Science, The University of Manchester Institute of Science and Technology, PO Box 88, Sackville Street, Manchester M60 1QD, UK.

Chapter 1

CYCLOPOLYMERISATION AND CYCLOCOPOLYMERISATION

G. C. CORFIELD

Sheffield City Polytechnic, UK

and

G. B. BUTLER

University of Florida, USA

SUMMARY

This review is concerned with chain-growth polymerisation reactions which lead to polymers containing cyclic units. Recent investigations of monomers, or combinations of monomers, which undergo cyclopolymerisation, or cyclocopolymerisation, are reviewed. A brief account is included of the properties of the commercially important poly(diallylamines) and the DIVEMA cyclocopolymer, which exhibits biological activity.

Considerable emphasis is placed upon studies related to the micro-structures of cyclopolymers and cyclocopolymers. Information has been gained from model radical cyclisation reactions and addition reactions of non-conjugated dienes. Direct analysis of polymers using modern spectro-scopic techniques, particularly ^{13}C n.m.r., has yielded significant results.

Theoretical interpretations of the driving force for these reactions are presented. The importance of kinetic versus thermodynamic control, electronic interactions, steric effects and charge-transfer complexes is discussed.

1. INTRODUCTION

It was during work reported in a series of papers published from 1949 to 1957 that Butler and coworkers found that several polymers produced from diallyl quaternary ammonium salts were soluble in water, and therefore not cross-linked, and yet they did not contain residual unsaturation. It was realised that, in these cases, both allyl double bonds were involved in a reaction leading to a linear polymer, and a mechanism was proposed in which chain growth occurred via alternating intra-molecular and inter-molecular steps (eqn. 1).[1] The presence of cyclic units in the main chain of these polymers was established by degradation reactions carried out on representative polymers.[2] In this mechanism, now known as *cyclopolymerisation*, cyclisation occurs as a characteristic feature of the polymerisation process and not as a deficiency in cross-linking and network formation.

After the cyclopolymerisation mechanism had been discovered, it became clear that a similar explanation could be offered for an unusual copolymerisation reaction. In 1951, Butler had synthesised a soluble 1:2 copolymer of divinyl ether and maleic anhydride which did not contain any carbon–carbon unsaturation. A mechanism (eqn. 2) now known as *cyclocopolymerisation*, leading to polymers containing bicyclic units, was proposed and supported by experimental evidence.[3]

Following these investigations, a vast amount of work has been carried out on cyclopolymerisation and cyclocopolymerisation; our earlier review cited more than 500 references[4] and a number of other reviews have since appeared.[5-7] Much of the earlier work went into the synthesis of new monomers and polymers to discover the range of structures which could be achieved and the properties that they possessed. In this review, we shall summarise the reactions which are encompassed within the terms cyclopolymerisation and cyclocopolymerisation and discuss the more recent investigations of new monomers or combinations of monomers which lead, by a chain-growth polymerisation mechanism, to polymers having cyclic units as the dominant feature of the polymer structure. Useful polymers have resulted from research in this field and a brief review of the properties of polymers of industrial and medical importance is included. However, particular emphasis will be placed upon those studies which have provided an analysis of the microstructure of the polymer chain and work which has attempted to explain the mechanism of the cyclisation reaction.

Although it has long been understood that properties are closely related to the structure of polymers, early workers in this field were often satisfied

$$(1)$$

when the linear nature of the polymers and the presence of cyclic units had been established. Thermodynamic factors were considered most important and hence structures **I** (eqn. 1) and **II** (eqn. 2) were assumed, though structures **III** and **IV**, respectively, are also feasible. Structure **I** follows from the expectation that the reaction would proceed via head-to-tail propagation, involving secondary radicals as intermediates, to produce a

$$(2)$$

six-membered ring (eqn. 1), whereas, structure **III** requires a head-to-head cyclisation step with formation of a primary radical, commonly accepted as a less stable intermediate, which leads to the more strained five-membered ring (see eqn. 16). Similar considerations apply to the formation of structures **II** and **IV**. Recent studies using a combination of chemical and spectroscopic methods (particularly ^{13}C n.m.r.) have yielded significant information on the structures of cyclopolymers. Other approaches to a study of the ring sizes of cyclic units in these polymers are to detect or trap the initially formed reactive species. Such structural studies are developing our understanding of the driving force for the cyclisation reaction.

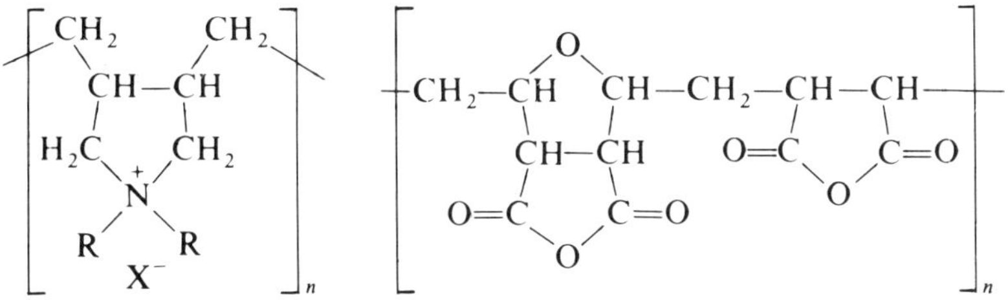

III

IV

Various attempts have been made to provide a general explanation for the characteristic features of the cyclopolymerisation and cyclocopolymerisation reactions. As a fundamental explanation of cyclopolymerisation, Butler proposed[8] that an electronic interaction occurs between the unconjugated double bonds of 1,6-dienes, or between the intramolecular double bond and the reactive centre, which provides a favourable pathway for cyclisation. Alternatively, Gibbs and Barton[9] considered steric effects to be more important. The presence of the large pendant group was regarded as a hindrance to intermolecular reaction and conformational interconversions would bring the unconjugated double bond to a favourable position for reaction. Later, unsatisfactory correlations between polymer structures and telomerisation reactions[10] led to concern over the relative importance of kinetic and thermodynamic factors in the cyclisation step.[5] In the cyclocopolymerisation process, participation of charge-transfer complexes has been proposed by Butler.[11] Studies of these ideas have provided information of importance not just to the mechanisms of cyclopolymerisation and cyclocopolymerisation but also to chain-growth polymerisation in general.

2. SYNTHESIS AND PROPERTIES

Our earlier review[4] demonstrated the scope of the cyclopolymerisation and cyclocopolymerisation mechanisms as a means of synthesising linear polymers having cyclic units in the main chain. We shall adopt the classification scheme used there but will present here a selective account of recent research in this field. Polymers with valuable properties have been discovered as a result of this research and a brief review of the properties of those of industrial and medical importance is included.

2.1. Synthesis of Cyclopolymers and Cyclocopolymers

2.1.1. Cyclopolymerisation of Symmetrical 1,6-Dienes

Since Butler's initial investigations, it has been shown[4,10] that a wide variety of 1,6-dienes can be polymerised to soluble, saturated polymers, and structures having five- and/or six-membered rings have been proposed (eqn. 3).

$$(3)$$

Cyclopolymerisation of diallylamines has been reviewed by several authors[6,12,13] and the papers presented at two symposia on this topic have been published.[14] Hawthorne and Solomon[15] have studied the effect of β-substituents on the cyclopolymerisation of diallylamines (eqn. 4). Electron spin resonance studies and analysis of products[16,17] show that the piperidine ring content increases with the bulk of the β-substituent, which causes a small decrease in basicity of the polymers.

$$(4)$$

Hodgkin et al.[18-21] have prepared a large number of diallylamines with a variety of polar and potential metal chelating groups as substituents on the nitrogen atom. Their ability to form cyclopolymers and the structures and some properties of these polymers were investigated. For cross-linking these polyamines, it has been shown[22] that use of 1,6-bis(diallylamino)-hexane (V) allowed optimum control of cross-linking, pK_a values and ion-exchange capacity.

$$N(CH_2)_6N$$

V

Jackson[23-25] has investigated the cyclopolymerisation of various diallylamines and has prepared thermally regenerable ion-exchange resins. He also reports on methods for the determination of residual unsaturation. Other workers[26-28] have studied the cyclopolymerisation of dialkyldiallyl-ammonium halides.

Danielyan *et al.*[29] observed that during the radical cyclopolymerisation of diallylcyanamide (**VI**), the presence of zinc chloride increased the polymerisation rate and polymer yield.

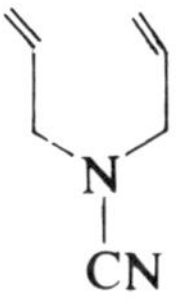

$$\text{VI}$$

Kodaira[30-32] has studied the radiation-induced cyclopolymerisation of N-substituted dimethacrylamides (**VII**), in the liquid, supercooled-liquid, glassy and crystalline states. For **VIIa-d**, cyclopolymerisation occurred completely, predominantly to five-membered rings, in all states except the glassy state, where no polymerisation was observed. The proportion of six-membered rings was a little higher in the crystalline state than in liquid or supercooled-liquid states. The cyclopolymerisation of **VIIe** showed some differences, which were attributed to the ordering effect of the octadecyl group.

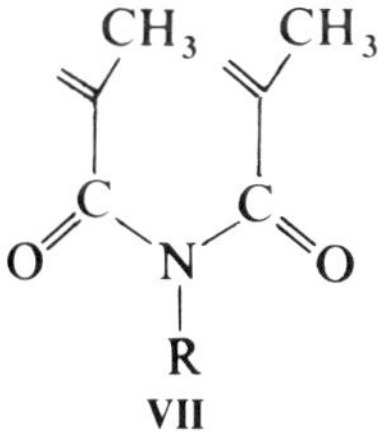

$$\text{VII}$$

a, $R = CH_3$
b, $R = C_6H_5$
c, $R = CH_3CH_2CH_2$
d, $R = C_6H_5CH_2$
e, $R = CH_3(CH_2)_{16}CH_2$

The cyclopolymerisation of divinylformal (**VIII**) has been investigated further. Yamakita and Hayakawa[33] have shown that gamma-ray bulk polymerisation in the solid state at $-190\,°C$ and $-78\,°C$ gave polymers having a linear structure with pendant vinyl groups, while polymers obtained at $0\,°C$, in the liquid state, or at $-78\,°C$ by solution polymerisation, contained cyclic units. Tsukino and Kunitake[34] have used ^{13}C n.m.r. spectroscopy to show that polymers of divinylformal contain a cis-4,5-disubstituted-1,3-dioxolane ring as the predominant unit, both in the main chain and as a pendant group (eqn. 5).

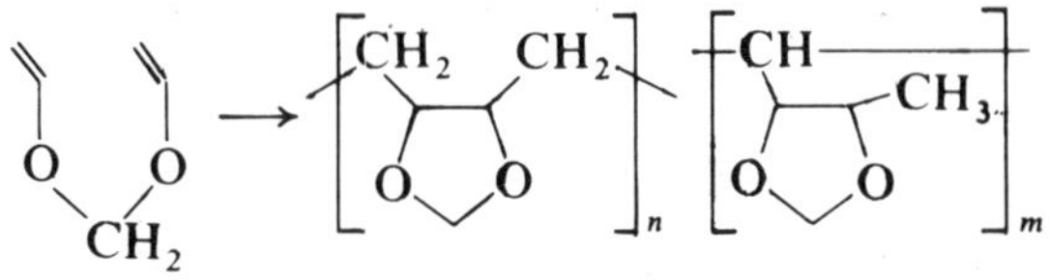

$$\text{(5)}$$

VIII

Kida *et al.*[35] have studied the polymerisation of methylvinyloxy-germanes (**IX**) using radical and cationic initiators. Di- and trivinyl-oxygermane radical polymerisations involved the formation of five- and six-membered rings. A number of diallylsilanes (**X**) have been prepared by Billingham *et al.* and their cyclopolymerisation reported.[36] In contrast to earlier work, soluble, fusible, linear cyclopolymers were obtained by radical polymerisation of divinyl phosphonates (**XI**) in dilute solutions, especially if the polymerisations were carried out at low temperatures and the initiator was decomposed photolytically.[37]

$$Me_{4-n}Ge(OCH{=}CH_2)_n$$

$$n = 1\text{--}3$$

IX

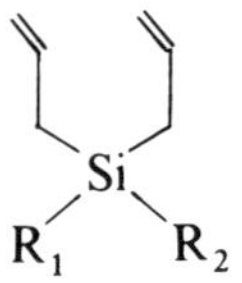

X

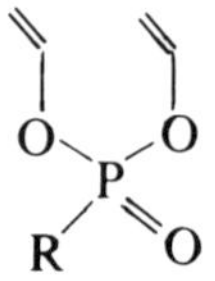

$$R = Me, Ph$$

XI

The polymerisation of 1,1'-divinylferrocene (**XII**) was first investigated by two independent research groups[38,39] who reported that soluble polymers could be obtained, using radical and cationic initiators, with the formation of three-carbon bridged ferrocene units in the chain (eqn. 6). Recent investigations of these polymers by Mössbauer spectroscopy[40] have shown that the radical- and cationic-initiated polymers have significantly different parameters (Table 1). Using model compounds it has been shown that radical initiation does yield cyclopolymers with the three-carbon bridged ferrocene unit but that polymers produced by cationic initiation do not have this structure.

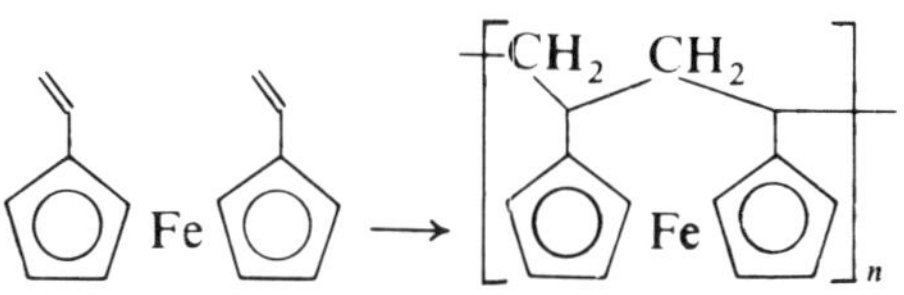

$$\text{(6)}$$

XII

TABLE 1
MÖSSBAUER PARAMETERS OF POLYMERS OF 1,1′-DIVINYL-
FERROCENE[40]

Initiator	Isomer shift[a] δ (mm s^{-1})	Quadrupole splitting ΔE_Q (mm s^{-1})
Radical	0·23 (2)	2·29 (2)
Cationic	0·27 (2)	2·40 (2)

[a] Relative to Fe in Pd.

2.1.2. Cyclopolymerisation of Other Non-conjugated Dienes

Guaita et al.[41] have re-investigated the polymerisation of o-divinyl-benzene. The kinetics and energetics of the process strongly support a proposal that the polymers are mainly comprised of seven-membered rings (**XIII**).

Large rings have been synthesised from a range of other monomers. Nishimura has reviewed his work on the polymerisation of 1,3-bis(4-vinylphenyl)propane and similar compounds.[42] Cyclopolymers having [3.3]paracyclophane repeating units (**XIV**) are only obtained by a cationic mechanism. Ethyleneglycol divinyl ether has been cyclopolymerised,[43] by radical and cationic initiators, to polymers having six (**XV**) and seven-membered rings (**XVI**), respectively, with some residual unsaturation (**XVII**). Diethyleneglycol divinyl ether,[44] by cationic initiation, yields a soluble polymer with ten-membered rings (**XVIII**), and tetraethyleneglycol divinyl ether gives a cross-linked polymer, claimed to contain 16-crown-5 units (**XIX**), which functions as a phase-transfer catalyst.[45] Butler has produced macrocyclic rings from a number of other monomers by cyclopolymerisation and cyclocopolymerisation.[46,47]

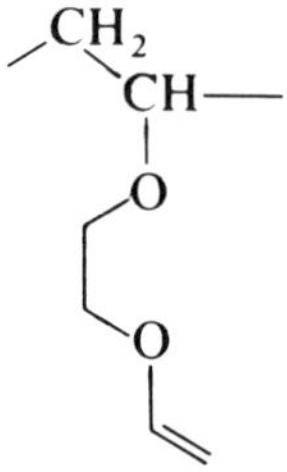

XVII 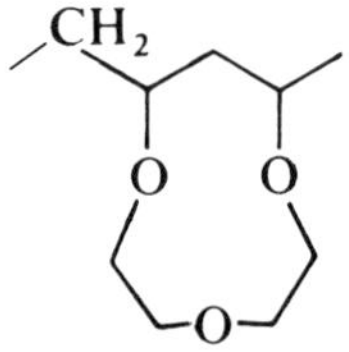   XVIII 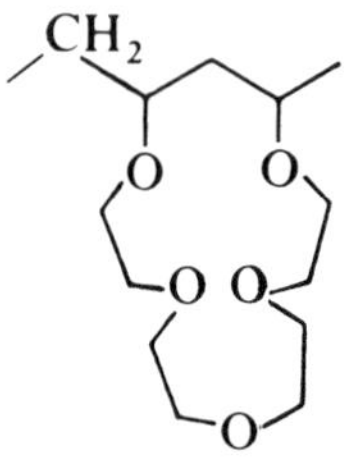   XIX

A number of dimethacrylates have been synthesised and cyclo-polymerised, including dialkylsilyl dimethacrylates.[48,49] Divinyl oxalate[50] and divinyl esters of other dibasic acids[51] have given polymers with cyclic structures. Hydrolysis to poly(vinyl alcohol) yields information on the propagation mechanism.

Matsumoto has continued his detailed studies of the polymerisation of diallyl compounds, particularly diallyl dicarboxylates.[52–54] Although poly(allyl propyl phthalate) contains about 15% of the head-to-head structure, on radical initiation at 80°C, polymers of diallyl phthalate contained lesser amounts, which were dependent upon the extent of cyclisation (Fig. 1). By extrapolation of results, completely cyclised poly(diallyl phthalate) was found to consist exclusively of an eleven-membered ring (**XX**) from head-to-tail propagation. This is not the case with polymers from diallyl *cis*-cyclohexane-1,2-dicarboxylate or diallyl succinate, where at first a decrease in head-to-head structures was noted but

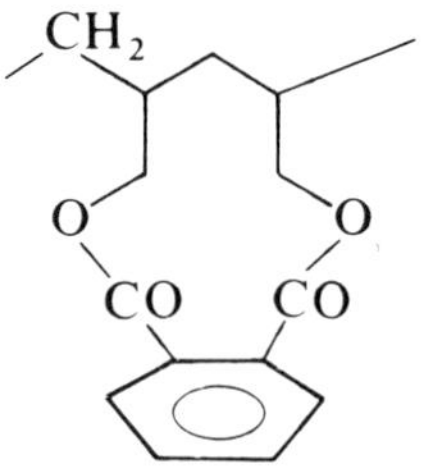

XX

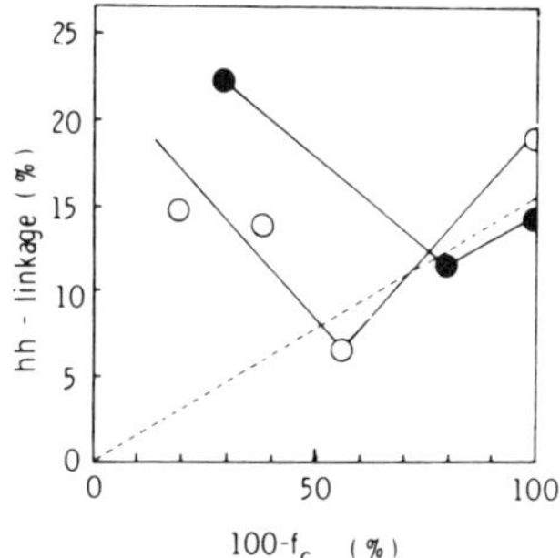

FIG. 1. Relationships between head-to-head (hh) structures and fraction of cyclisation (f_c) for diallyl *cis*-cyclohexane-1,2-dicarboxylate (○), diallyl succinate (●) and diallyl phthalate (– – – –). (Reproduced from Matsumoto, A., Iwanami, K. and Oiwa, M., *J. Polym. Sci., Polym. Chem. Ed.*, **19** (1981), p. 214, by permission of the publishers, John Wiley & Sons, Inc. ©.)

then an increase occurred at high extents of cyclisation' (Fig. 1). An explanation based on steric factors is provided. Allyl esters of unsaturated acids have also been studied in detail by Matsumoto.[55]

Yokota and Takada[56] have polymerised *o*-isoprenylphenyl vinyl ether (**XXI**) by cationic and radical initiation to soluble cyclopolymers. The relative reactivities of the two functional groups in cationic polymerisation were estimated from the copolymerisation of phenyl vinyl ether and α-methylstyrene. During cyclopolymerisation, inter-molecular addition of a vinyl ether group was always followed by cyclisation, but addition to an isoprenyl group led to cyclisation or further inter-molecular reactions. Hence only vinyl ether groups are present in any residual unsaturation in the polymers. Polymers produced by radical initiation in bulk are completely cyclised. Infrared spectroscopy showed that cyclisation occurred via head-to-head addition, leading to five-membered rings (eqn. 7).

Yokota *et al.*[57–62] have also studied the cyclopolymerisation of a

(7)

number of other unsymmetrical monomers, among which are vinyl *o*-isoprenyl benzoate, *o*-allylphenyl acrylate, *o*-vinylphenyl acrylate and 2-(*o*-allylphenoxy)ethyl acrylate.

Cyclopolymerisation of *N*-allyl-*N*-methylmethacrylamide (**XXII**) has been investigated by Kodaira *et al.*[63,64] Radical initiation over the temperature range -78 to $120\,^{\circ}\mathrm{C}$ resulted in polymers with 88–93 % cyclic units, mainly five-membered rings with some six-membered rings, and pendant methacryl groups (eqn. 8). Panzig and Mulvaney[65] have prepared polyampholytes by cyclopolymerisation of *N*-(2-phenylallyl)acrylamide and the *N*-ethyl derivative (**XXIII**), followed by hydrolysis of the cyclopolymer (eqn. 9). Cyclisation was not complete since cross-linking readily occurred.

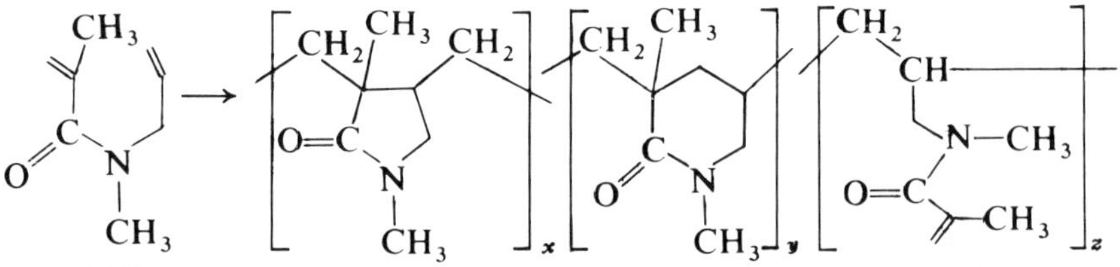

$$\textbf{XXII} \tag{8}$$

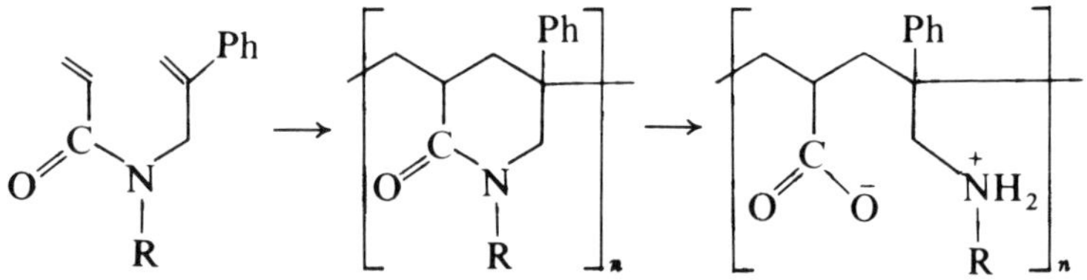

$$\textbf{XXIII} \quad R = H, Et \tag{9}$$

2.1.3. Cyclopolymerisation Leading to Polycyclic Systems

The earlier reports of the formation of cyclopolymers containing polycyclic structures by a succession of ring closures within a monomer[4] have been questioned by Solomon and Hawthorne.[6] The work of Beckwith *et al.*[66,67] has shown that the structure (**XXV**) which is assigned to cyclopolymers of triallyl monomers (**XXIV**) is likely to be incorrect, and (**XXVI**) and (**XXVII**), resulting from 1,5–1,6 and 1,5–1,7 cyclisations respectively, are more likely. Further, it is noted that there is a very much smaller rate constant for the second than for the first cyclisation, hence bicyclic units will only be favoured by very low monomer concentrations.

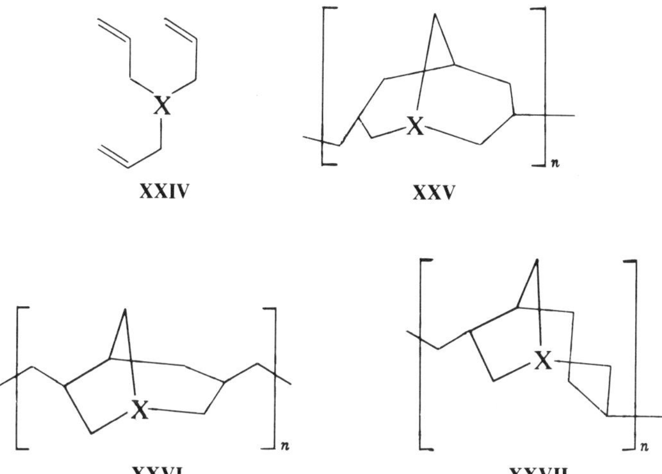

XXIV **XXV**

XXVI **XXVII**

At higher concentrations, cross-linking will occur or residual unsaturation will remain. It is suggested[6] that earlier workers reported high extents of cyclisation because of their inability to measure residual unsaturation accurately.

Polymers with bicyclic units formed by transannular polymerisations have been produced from a number of monomers. 4-Vinylcyclohexene[68,69] has been polymerised using various initiators and the three different structural units (**XXVIII–XXX**) quantified by infrared and n.m.r. spectroscopy.

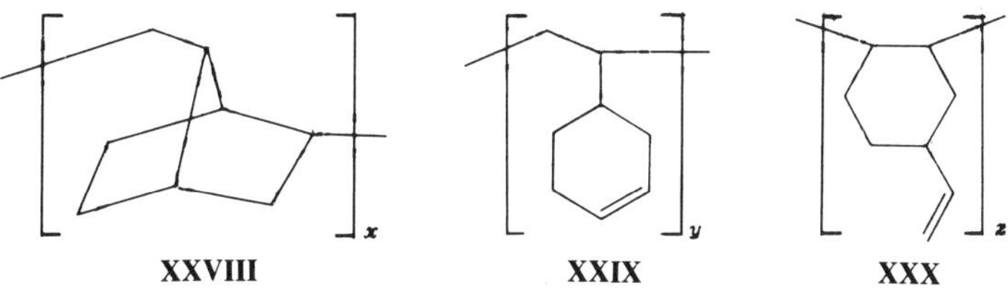

XXVIII **XXIX** **XXX**

Van Heiningen and Butler[70] have studied the cyclopolymerisation of 3-allylcyclopentene, 4-allylcyclopentene, 3-allylcyclohexene and 4-allylcyclo-hexene. Pinazzi *et al.*[71] have shown that cationic and Ziegler–Natta catalysed polymerisations of 1,2,6-cyclononatriene gave high yields of low molecular weight polymers containing a bicyclo[4.3.0]nonene structure (eqn. 10).

14 G. C. CORFIELD AND G. B. BUTLER

Polymerisation of 1,4-dienes can lead to bicyclic units which have formed from two molecules of the monomer.[4,10] Kunitake and Tsukino[72] have shown that radical polymerisation of divinyl ether yields partially cyclised

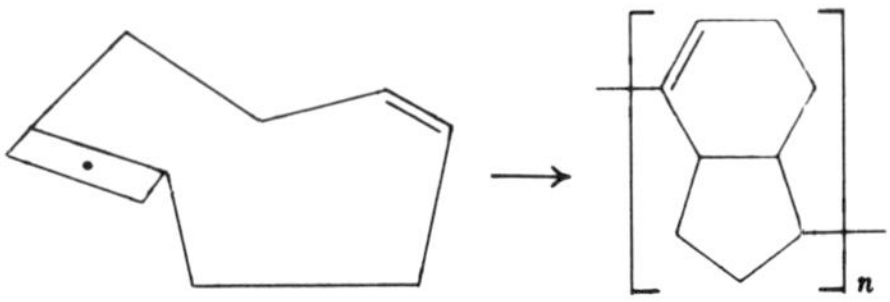

$$(10)$$

polymers, which, based on ^{13}C n.m.r. spectroscopy, contain a five-membered ring monocyclic unit (**XXXI**) and a dioxobicyclo[3.3.0]octane unit (**XXXII**) in a 1:1 ratio.

Guaita *et al.*[73] have investigated the relationship between polymer composition and monomer concentration for the radical polymerisation of divinyl ketone. Of the possible structural units, application of kinetic relationships provides evidence that six-membered monocyclic (**XXXIII**) and bicyclic rings (**XXXIV**) predominate, which is supported by infrared spectroscopy.

Recently the cyclopolymerisation of α,α'-dimethoxycarbonyldivinyl-amine was reported and a five-membered ring bicyclic structure proposed (eqn. 11) from spectral data.[74]

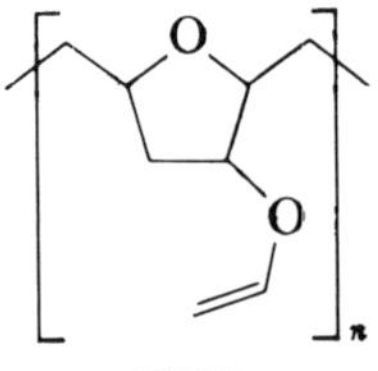

XXXI

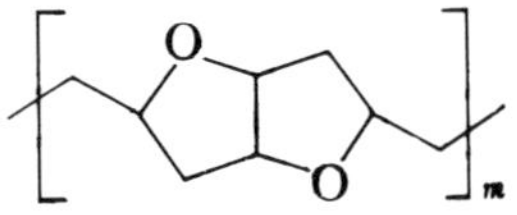

XXXII

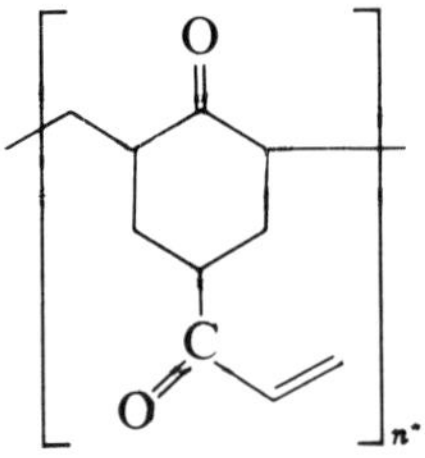

XXXIII

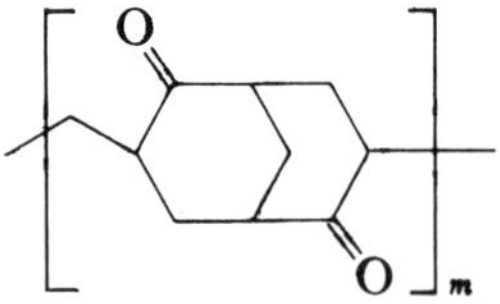

XXXIV

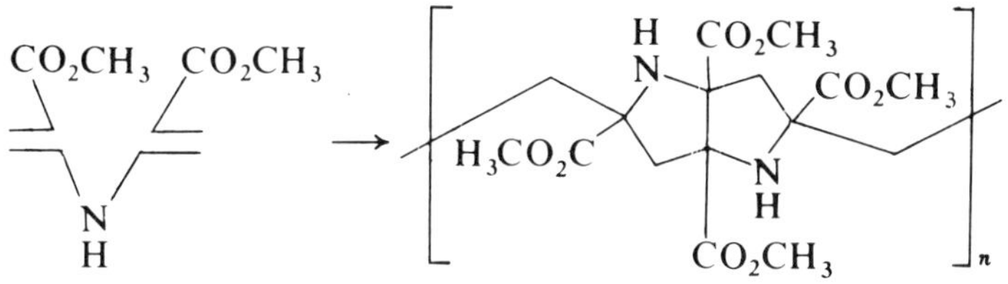

$$(11)$$

2.1.4. Cyclopolymerisation Involving Other Functional Groups

Diynes, diepoxides, dialdehydes, di-isocyanates and dinitriles have already been shown to undergo cyclopolymerisation.[4,10]

Matsoyan *et al.*[75-78] have investigated the polymerisation of a variety of diynes using palladium chloride as catalyst. Cyclopolymers with conjugation in the main chain have been obtained from monomers such as **XXXV** and **XXXVI**. Electrical conductivities and other properties of the polymers are reported. Recently, Gibson *et al.*[79-80] have reported the synthesis of the cyclopolymer of 1,6-heptadiyne as a free-standing film which can be doped with iodine to yield an electrically conducting polyene film.

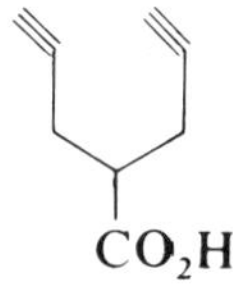

XXXV

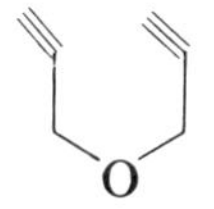

XXXVI

N,N-Bis(2,3-epoxypropyl)aniline (**XXXVII**) undergoes ring opening and cyclopolymerisation, to give linear polymers containing 2,4,6-substituted morpholindiyl units (eqn. 12), using potassium *tert*-butoxide as initiator.[81]

Kunitake[82] has continued his research on the polymerisation of aromatic aldehydes and observed that **XXXVIII** and **XXXIX** are partially cyclised by ionic initiators to polyacetals with seven-membered ring structures.

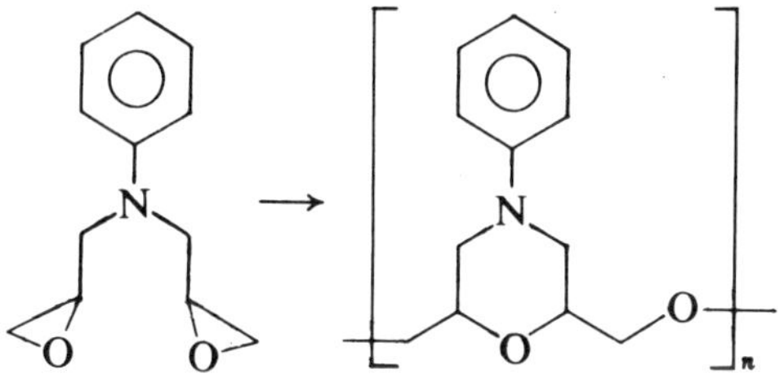

$$(12)$$

XXXVII

16   G. C. CORFIELD AND G. B. BUTLER

Bur and Fetters[83] have reviewed the literature on the cyclo-polymerisation of di-isocyanates. Woehrle[84] has polymerised succino-nitrile with cationic initiators to polymers, containing cyclic units (**XL**), which were semiconductors.

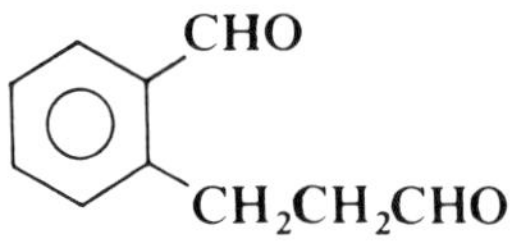

XXXVIII 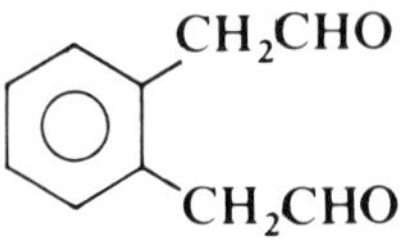 **XXXIX** 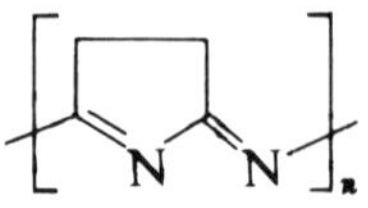 **XL**

2.1.5. Cyclocopolymerisation

Numerous examples of the copolymerisation of 1,6-, or other non-conjugated dienes, with vinyl monomers, or other co-monomers such as sulphur dioxide, are reported in the literature. In most cases the diene cyclopolymerises and soluble polymers are obtained. For example, Sayadyan *et al.*[85] have produced soluble copolymers of diallylcyanamide with vinyl chloride and vinyl acetate in which the diallyl monomer cyclises to *N*-cyanopiperidine units. Also, Amemiya *et al.*[86] have produced further cyclic copolymers of diallyl compounds and sulphur dioxide.

Cyclocopolymerisation[3,4,10] refers to copolymerisations where the cyclic repeating unit contains components of both monomers, as illustrated by the copolymerisation of divinyl ether and maleic anhydride (eqn. 2). Fujimori and Butler[87] have prepared cyclocopolymers of divinyl ether and tetrahydronaphthoquinone, and a derivative, in which structures **XLI** are found. These authors have also studied the cyclocopolymerisation of divinyl ether and several substituted maleic anhydrides which produced regular cyclocopolymers having 1:1 or 1:2 (divinyl ether:anhydride) compositions.[88] Acrylonitrile spontaneously copolymerises with divinyl ether, in the presence of zinc chloride, with a strong tendency to form a 1:2 alternating cyclocopolymer (**XLII**).[89] Butler[90] has recently reviewed the

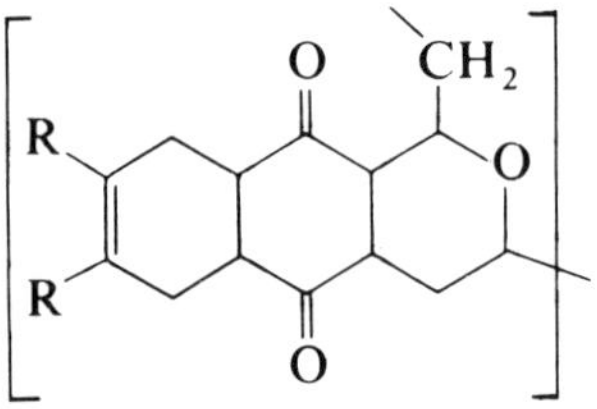

R = H, CH$_3$

XLI   **XLII**

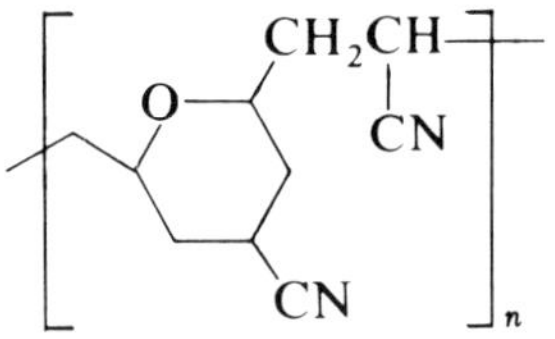

literature concerned with the synthesis and characterisation of cyclo-copolymers of this type.

Cyclocopolymers have been obtained also by copolymerisations of divinyl sulphone with ethyl vinyl ether or dihydropyran,[91] divinyl ether with furfural[92] and 1,6-heptadiene with sulphur dioxide.[93]

2.2. Properties of Cyclopolymers and Cyclocopolymers of Industrial and Medical Importance

2.2.1. Poly(diallylamines)

The first cyclopolymer to be manufactured in commercial quantities apparently was poly(dimethyldiallylammonium chloride).[94] This polymer, now manufactured by a number of suppliers, has been shown to possess optimum functional properties for application to electrographic paper reproduction processes.[95] In such processes, proper functioning of the photoresponsive coating depends on the rapid dissipation of static electrical charges by the substrate. Electroconductivity, normally expressed as resistivity, is dependent upon the water content in the substrate. Polyelectrolytes, because of their ionogenic nature, provide both a source of ions and a high degree of hygroscopicity; both properties are necessary in order to provide electroconductivity, a property believed to be ionic in mechanism rather than electronic, to the paper at low humidities, where the untreated sheet behaves as a dielectric. Poly(dimethyldiallylammonium chloride) was reported to give the best performance for this application among a variety of non-ionic, anionic, and cationic water-soluble polymers evaluated.

An applications analysis of cyclopolymers of N,N-dialkyldiallyl-ammonium halides has recently been published.[12] In addition to the use of these materials in the electroconductive paper reproduction process, other uses as paper additives include antistatic agents, fluorescent whiteners, paperboard reinforcement and retention aids. In the water treatment field, these polymers are used as flocculants and/or as primary coagulants or coagulant aids in potable water, waste water or sewage sludge treatments. They are also reported to be used in the zinc, tin and lead electroplating industries as well as in cosmetics, as a biocide in water, as a demulsifier of dispersed oils, and as a detergent additive. The copolymers of dialkyl-diallylammonium salts with sulphur dioxide first reported by Harada and Katayama[96] are manufactured commercially in Japan and have similar industrial uses and properties to the homopolymers reported above.

Although anion-exchange materials containing the quaternary am-monium cation in the polymeric network were synthesised as early as

1949,[97] a reinvestigation in this area[98] has shown that N,N,N',N'-tetraallyl-N,N'-dimethylethylenediammonium dichloride (**XLIII**) could be polymerised in suspension with ammonium persulphate as initiator to yield ion-exchangers for use in extraction of uranium of both superior rate and capacity in comparison with other exchangers tested.

$$\begin{array}{ccc} CH_3 & & CH_3 \\ | & & | \\ (CH_2{=}CHCH_2)_2\overset{+}{N}(CH_2)_2\,\overset{+}{N}(CH_2CH{=}CH_2)_2 \\ Cl^- & & Cl^- \end{array}$$

XLIII

Polytriallylamine, synthesised by polymerisation of triallylamine hydrochloride, either by use of conventional free-radical initiators or by irradiation with gamma-rays from a ^{60}Co source,[99] is reported to be superior to other weak-base amine polymers for use in the Sirotherm demineralisation process. This process, described in several publications,[100] is reported to be a thermally regenerated ion-exchange process for the cheap removal of salinity from brackish waters. The process involves the adsorption of salt from an aqueous solution in contact with a mixed-bed of a suitable pair of ion-exchange materials consisting of a carboxylic and a weak-base amine exchanger. The basis of the process which makes it workable is that weak-base amine exchangers in the salt form, and to a lesser extent carboxyl exchangers, show reduced ionisation on heating, thus reversing the adsorption process, and effecting regeneration of the mixed-bed ion-exchange system.

In a recent paper[101] it was shown that an amine exchanger prepared by copolymerising diallylamine hydrochloride with a suitable cross-linker gave superior thermally regenerable capacity in the 20–80 °C range to other amine exchangers evaluated. Butler[13] has recently discussed the properties of poly(diallylammonium salts).

2.2.2. *DIVEMA and Other Cyclocopolymers*

The use of synthetic polymers as biologically active materials is increasing. Among those polymers receiving most attention by investigators is the alternating cyclocopolymer of divinyl ether and maleic anhydride, commonly known as pyran copolymer or DIVEMA. This copolymer has been investigated extensively for a variety of biological effects since the initial observation by scientists at the National Cancer Institute that the hydrolysed and neutralised copolymer had considerable antitumour activity along with a much reduced toxicity in comparison with other polyanions investigated. As pointed out by Breslow,[102] DIVEMA has also been shown to be an interferon inducer, to possess antiviral, antibacterial, antifungal, anticoagulant and antiarthritic activity, to aid in removing polymeric plutonium from the liver, to inhibit viral RNA-dependent DNA polymerase (reverse transcriptase), and to activate macrophages in affecting the immune response of test animals. The interferon-inducing capability was initially observed during a clinical investigation of DIVEMA as an antitumour drug. In a recent review[90] the importance of molecular weight and molecular weight distribution was discussed. A biphasic response of the reticuloendothelial system to the copolymer drug had been observed which led to the postulate that the copolymer may consist of a toxic molecular weight fraction, and another fraction of lower toxicity. Other evidence for molecular weight dependence is based upon the observation that antiviral activity of the copolymer requires a higher molecular weight than for antitumour activity. The latter was optimum with low molecular weight samples of narrow molecular weight distribution. These results led to development of a method for synthesising a copolymer of narrow molecular weight distribution, by photochemically initiating solution polymerisation in acetone, with tetrahydrofuran as a chain-transfer agent. This method gave samples having a molecular weight distribution ($\bar{M}_w / \bar{M}_n$) ranging from 1·6 to 2·6, in contrast to 3·7 to 7·7 for earlier prepared samples.

Breslow[103] has discussed recent progress in the use of DIVEMA and Butler[90] has reviewed the biological activity of polymers structurally related to pyran copolymer.

3. MICROSTRUCTURE OF POLYMERS

3.1. Initial Investigations

3.1.1. Cyclopolymers

The mechanism proposed by Butler and Angelo[1] to explain the formation of soluble, and hence linear, saturated polymers from diallyl quaternary ammonium salts assumed that six-membered ring cyclic units would be produced (**I**, eqn. 1). To confirm the proposed structure, representative polymers were degraded, by methods which could cleave the cyclic units in the chain, and the expected products were obtained (eqns. 13 and 14). Again, it was assumed that the polymers contained a six-membered ring.[2]

This was a natural assumption in view of the thermodynamic stability of six-membered rings and the fact that the intermediates involved would be secondary radicals arising from a head-to-tail chain propagation reaction. Marvel and co-workers[104] similarly proposed six-membered cyclic structures for the polymers from 1,6-heptadiene and related compounds on the evidence that partial dehydrogenation of the polymers gave products whose infrared and ultraviolet spectra were characteristic of *meta*-substituted aromatic rings. In agreement with this work, Friedlander[105] reported the exclusive formation of six-membered cyclic monomeric products in the free-radical addition reactions of 1,6-dienes with various addenda (eqn. 15, for example).

$$(13)$$

$$(14)$$

As a result of these studies, the structures of most new cyclopolymers from 1,6-dienes were given six-membered ring units. In many cases no proof of structure was given or other ring size considered.[4,5,10] Since initiation and propagation steps could occur by addition to either the *head* or *tail* of the double bonds, there exists the possibility for formation of five-, six- or even seven-membered ring structures (eqn. 16).

Early evidence for the existence of structures other than a six-membered ring came from Arbuzova and Sultanov,[106] who found that the polymerisation of divinyl acetals gave polymers containing both five- and six-membered rings (eqn. 17). Determination of the distribution of cyclic

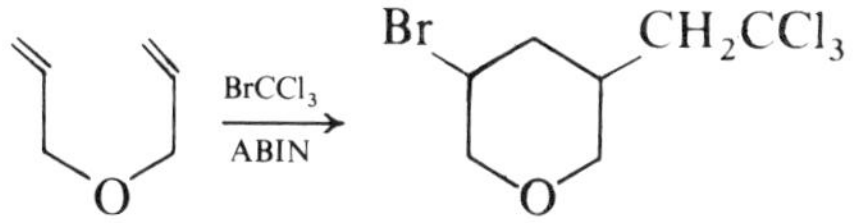

$$(15)$$

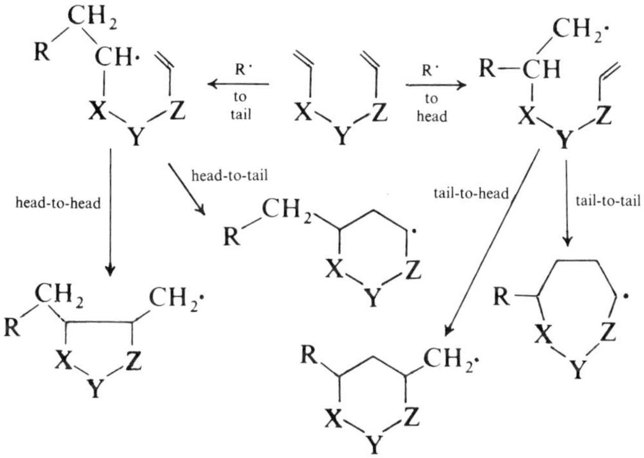

$$(16)$$

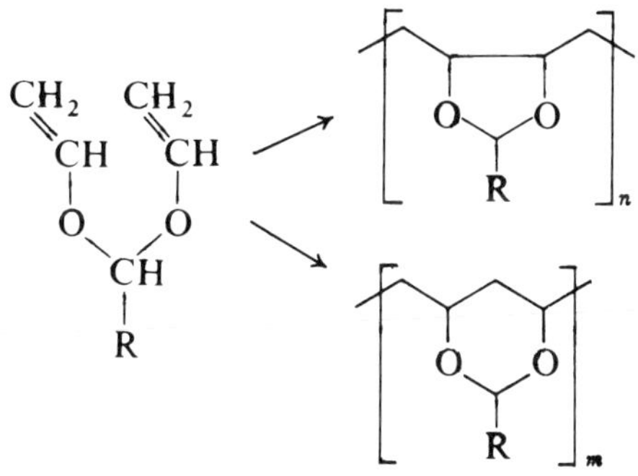

$$(17)$$

units was carried out by hydrolysis of the cyclopolymer followed by analysis of the derived poly(vinyl alcohol) for 1,2- and 1,3-glycol content. Matsoyan[107] used this technique to determine the ring content of a large number of poly(divinyl acetals). The results gave $n:m = 23:77$ (eqn. 17) and were independent of R. Murahashi *et al.*[108] determined the 1,2-glycol content in poly(vinyl alcohol) derived from divinyl carbonate and values equivalent to 6–23 % of five-membered rings were found.

Other types of cyclopolymerisation reactions were also found to yield five-membered rings. A polymer with a relatively high content of five-membered cyclic anhydride units was obtained by polymerisation of acrylic anhydride at 115 °C in xylene.[109] Also, differences were observed in the carbonyl region of the infrared spectra between cyclopolymers from dimethacrylamides and polymers produced by deamination of polymethacrylamides.[110] It was concluded that the cyclopolymerisation of dimethacrylamides gave predominantly five-membered ring units (eqn. 18).

The question of the ring size produced in radical addition reactions was reconsidered after Lamb *et al.*[111] observed predominantly five-membered rings in the thermal decomposition of 6-heptenoyl peroxide (eqn. 19).

$$R = H, Me, Et, Pr \text{ or } Ph \tag{18}$$

$$\tag{19}$$

94 %

Brace[112] found that in the cyclisation of 1,6-heptadiene with bromo-trichloromethane and other addenda, five-membered ring compounds were obtained, results which contradicted Friedlander's earlier work. He extended his study to include diallyl ether, ethyl diallylacetate and diallylcyanamide. In each case predominantly five-membered ring products were found (eqn. 20, for example). Aso[113] also observed five-membered ring products from the cyclisation of divinyl formal.

Five-membered rings resulted from radical cyclisation reactions of 6-bromo-1-hexene and other alkenyl halides[114] (eqn. 21). In the cyclisation of allyl ethenesulphonate with butyl mercaptan[115] only the five-membered ring sultone and sulphonium salt were observed (eqn. 22). Previously, in the cyclopolymerisation of this monomer, the same authors had proposed only six-membered rings.[116]

$$\text{(20)}$$

$$\text{(21)}$$

$$\text{(22)}$$

3.1.2. Cyclocopolymers

The structure proposed for the cyclocopolymer of divinyl ether and maleic anhydride (**II**, eqn. 2) was supported[3] by: (i) an elemental analysis in agreement with a 1:2 molar ratio of divinyl ether:maleic anhydride; (ii) incorporation of a small amount of iodine in the structure after reaction with hydriodic acid, a result interpreted as due to cleavage of some cyclic ether units; (iii) a broad infrared absorption band at $1085–1100\,\text{cm}^{-1}$, attributed to the C—O str. of a *six*-membered cyclic ether, in addition to bands characteristic of a cyclic anhydride. However, it is difficult to identify an unknown cyclic ether using the C—O str. since all ethers have strong broad absorptions in this region. Tetrahydrofuran has an intense broad absorption centred at $1070 \, cm^{-1}$ and tetrahydropyran has numerous strong bands covering the region $1000-1100 \, cm^{-1}$. Therefore, the presence of a polymer unit involving a five-membered cyclic ether (**IV**) should not be ruled out.

Support for the six-membered ring unit in a cyclocopolymer of divinyl ether and dichloromaleic anhydride has been claimed[87] on the basis of reaction of the polymer with sodium hydroxide which was followed by back titration using high-frequency oscillometry. It was suggested that only 1 mole of hydrogen chloride was eliminated. Further, it was proposed that only a six-membered cyclic ether unit (**XLIV**) could lose hydrogen chloride and that it would be 1 mole only (eqn. 23). Loss of a second mole, it is stated, would be extremely difficult due to the destabilising effect of the adjacent oxygen atom on the incipient negative charge. In the five-membered ring unit (**XLV**, eqn. 24) both hydrogens which are to be lost would be on a carbon atom adjacent to the oxygen atom. However, it has been shown that thermal decomposition of 3-oxa-6-azoni-aspiro[5,5]undecane hydroxide (eqn. 25) gave a mixture consisting of

$$\text{XLIV} \xrightarrow[-\text{HCl}]{\text{aq. NaOH}} \qquad \qquad (23)$$

$$\text{XLV} \xrightarrow{\quad\times\quad} \qquad \qquad (24)$$

$$\text{XLVI} \longleftarrow \quad \bar{O}H \longrightarrow \text{XLVII} \qquad \qquad (25)$$

$$86\% \qquad\qquad\qquad\qquad 14\%$$

1-(2-vinyloxyethyl)piperidine (**XLVI**, 86 %) and 4-(pent-4-enyl)morpholine (**XLVII**, 14%). Thus, elimination in the morpholine ring is appreciably easier than in the piperidine ring.[117] This suggests that the inductive effect of the oxygen atom causes an increased acidity of the adjacent C—H bond. Without further evidence, particularly studies on model compounds, it is unwise to rule out the existence of a five-membered ring structure in the cyclocopolymer of divinyl ether and dichloromaleic anhydride.

3.2. Recent Developments

It has become clear that the structures of cyclopolymers or cyclocopolymers cannot be assumed to contain only the thermodynamically most stable rings. In recent years, it has been realised that a detailed analysis of the microstructure of the polymer chain is necessary before the mechanism of the cyclisation reaction and the properties of the polymers can be understood. There have been three approaches to this problem:

(a) a study of intra-molecular addition reactions of alkenyl radicals;
(b) detection or trapping of low molecular weight products from addition reactions of non-conjugated dienes;
(c) direct investigation of polymers using modern spectroscopic methods.

3.2.1. Radical Cyclisation

The question being asked is illustrated in eqn. 26. If a free radical is generated within a structure in which there is an accessible unsaturated centre does *endo-* or *exo*-cyclisation occur? Clearly, this is a model situation for the cyclopolymerisation reaction.

A large number of reactions of this type have been investigated, some of which have been mentioned already,[111,114] and several reviews have been published.[118-120] Recently, Beckwith *et al.*[121] have presented some generalisations concerning the influence of steric and stereo-electronic

$$\text{(26)}$$

factors on radical reactions of this type. The following guidelines are suggested:

(i)　intra-molecular addition under kinetic control in lower alkenyl radicals and related structures occurs preferentially in the *exo*-mode;

(ii)　substituents on an olefinic bond disfavour homolytic addition at the substituted position;

(iii)　1,5-ring closures of substituted hex-5-enyl and related radicals are stereoselective.

Examples[122,123] of guideline (i) include hexenyl radicals (**XLVIII**; R_1–R_5 = H) and variously substituted hexenyl radicals (for example **XLVIII**; R_1, R_2 = Me, R_3–R_5 = H) and the 2-allyloxyethyl radical (**XLIX**; R = H).[114] In these cases the reactions are irreversible and the major

$$R_5$$
$$R_4$$
$$R_1 \quad R_2 \quad R_3$$

XLVIII　　　　　**XLIX**

products, by far, are five-membered ring structures arising from *exo*-cyclisation. However, in other cases (for example **XLVIII**; R_1 = CN, R_2 = CO$_2$Et, R_3–R_5 = H) where the initial radical is stabilised by neighbouring groups, the reactions are reversible and under thermodynamic control which leads to the six-membered ring product by *endo*-cyclisation. *Exo*-ring closure of radicals of this type is the kinetically favoured process because the stereo-electronic requirements of the transition state can be most readily satisfied. The addition of a free radical to an alkene will occur most readily when the orbitals bearing the three electrons which are redistributed remain in the same plane throughout the reaction.[124,125]

The 5-substituted hex-5-enyl radicals (**XLVIII**; R_3 = Me, R_1 = R_2 = R_4 = R_5 = H) and (**XLVIII**; R_3 = Pr^i, R_1 = R_2 = R_4 = R_5 = H) illustrate guideline (ii). These have been found to undergo mainly six-membered ring closure[123,126] because the rate of cyclisation to five-membered rings is greatly retarded due to the presence of the substituent. However, conflicting results have been observed for the 5-methylhex-5-enyl[127] and 2-methallyloxyethyl radicals (**XLIX**; R = Me),[128] where five-membered rings were still found to predominate. With the 2-(2-phenylallyloxy)ethyl radical (**XLIX**; R = Ph)[128] six-membered rings do predominate, but here the intermediate radical is considerably stabilised by resonance.

Cyclisation of 1- or 3-substituted hex-5-enyl and related radicals occurs regiospecifically in the *exo*-mode and yields mainly *cis*-disubstituted cyclic products (eqn. 27), whereas 2- or 4-substituted systems give mainly *trans*-products (eqn. 28),[129–131] illustrating guideline (iii). The formation of the *cis*-isomer from 1-substituted hexenyl radicals has been attributed to orbital symmetry control.[129] The ring closure of 2-, 3-, or 4-substituted hexenyl radicals is related to the conformational preference of the transition states.[130]

$$\text{(27)}$$

$$\text{(28)}$$

In these models of the cyclopolymerisation reaction, significant information concerning the cyclisation of radicals has been discovered. But are similar products obtained from radical addition reactions of non-conjugated dienes?

3.2.2. Addition to Non-conjugated Dienes

Beckwith *et al.*[66,132,133] have used e.s.r. spectroscopy to study the radical addition reactions of non-conjugated dienes. Hydroxyl, amino, methyl or phenyl radicals were reacted with diallylmalonic acid (and its derivatives), diallyl ether, diallylamine and variously substituted diallylamines in aqueous solutions. Using a flow technique, the products of the reaction were generated in the cavity of an e.s.r. spectrometer. The results were interpreted as clear evidence that the radicals present in highest concentration in the flow cell when 1,6-dienes of general type **L** react are five-membered ring radicals. For example, the spectrum obtained from the reaction of diallylamine with hydroxyl radicals (Fig. 2) consists of a triplet (splitting due to α protons) of doublets (splitting due to β protons). This

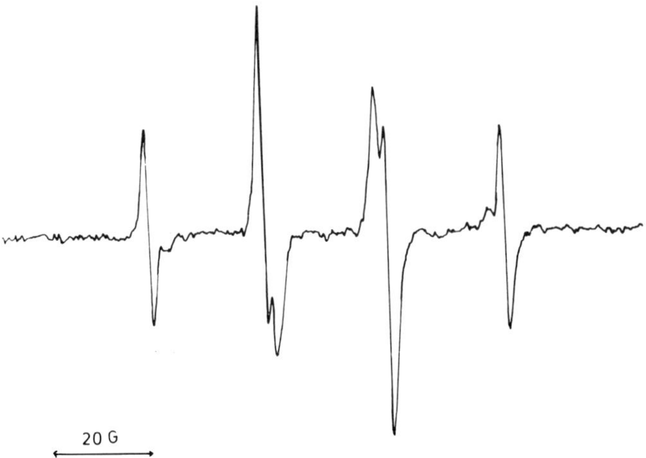

FIG. 2. Electron spin resonance spectrum of radicals formed from diallylamine and hydroxyl radicals. (Reproduced from Beckwith, A. L. J., Ong, A. K. and Solomon, D. H., *J. Macromol. Sci., Chem.*, **A9** (1975), p. 127, by permission of the publishers, Marcel Dekker, Inc. ©.)

spectrum is assigned to the five-membered ring radical (**LI**), rather than the six-membered ring radical (**LII**) or the uncyclised radicals (**LIII** and **LIV**), since it requires that the unpaired electron interacts with three protons, two of which are equivalent. Only **LI** and **LIV** have such a structure. The uncyclised radical (**LIV**) is rejected since this constitutes an abnormal mode of addition to an allylic double bond.

The sensitivity of this cyclisation to steric and resonance effects was shown by investigations on a number of substituted diallylamines. The e.s.r. spectrum of radicals from N,N-bis(2-isopropylallyl)methylamine and $NH_3^{+\cdot}$ (Fig. 3) shows a multiplet spectrum which is assigned to six-membered ring radicals. This is in accord with other results on cyclisation of radicals[123] and shows that a bulky substituent on the internal position of the double bond disfavours attack at that position.

Solomon and co-workers[15,16] have investigated the products from the reactions of diallylamine derivatives with radicals. Conditions were chosen such that chain propagation to high molecular weight did not occur and the low molecular weight products were isolated and characterised by mass spectrometry and ^{13}C n.m.r. spectroscopy.[15,16,134,135] N,N-Diallyl-methylamine and derivatives have been studied in most detail and the results are summarised in Table 2 and eqn. 29. After reacting the amines, hydrogen chloride and azobisisobutyronitrile in ethanol at 60 °C, the products were isolated and characterised. Identified among the basic products were perhydroisoindol-5-ones (**LV**), 3-azabicyclo[3.3.1]nonan-6-imines (**LVI**), or a mixture of these compounds. In certain cases **LVIII** and

TABLE 2

RADICAL ADDITION^a TO N,N-DIALLYLMETHYLAMINES (**LVII**)

Reactant (**LVII**)		Products^b	
R_1	R_2	(%)	Structure
H	H	100	**LV** ($R_1 = R_2 = H$)
CH_3	H	59	**LV** ($R_1 = CH_3$, $R_2 = H$)
		41	**LV** ($R_1 = H$, $R_2 = CH_3$)
CH_3	CH_3	77	**LV** ($R_1 = R_2 = CH_3$)
		22	**LVI** ($R_1 = R_2 = CH_3$)
C_2H_5	H	45	**LV** ($R_1 = C_2H_5$, $R_2 = H$)
		55	**LV** ($R_1 = H$, $R_2 = C_2H_5$)
C_2H_5	C_2H_5	100	**LVI** ($R_1 = R_2 = C_2H_5$)
$(CH_3)_2CH$	H	25	**LV** ($R_1 = (CH_3)_2CH$, $R_2 = H$)
		75	**LV** ($R_1 = H$, $R_2 = (CH_3)_2CH$)
$(CH_3)_2CH$	$(CH_3)_2CH$	100	**LVI** ($R_1 = R_2 = (CH_3)_2CH$)
$(CH_3)_3C$	H	25	**LV** ($R_1 = (CH_3)_3C$, $R_2 = H$)
		25	**LV** ($R_1 = H$, $R_2 = (CH_3)_3C$)
		50	**LVIII**
C_6H_5	H	50	**LV** ($R_1 = C_6H_5$, $R_2 = H$)
		20	**LIX**

^{a} Cyanoisopropyl radical in ethanol at 60 °C.
^{b} Determined by mass and ^{13}C n.m.r. spectroscopy.

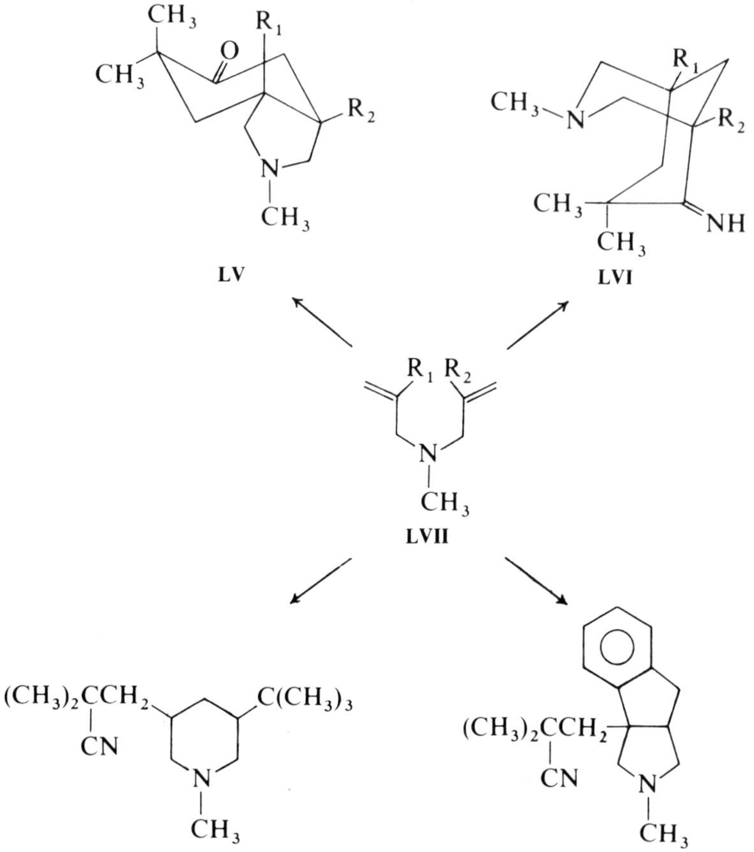

LV

LVI

LVII

LVIII

LIX

(29)

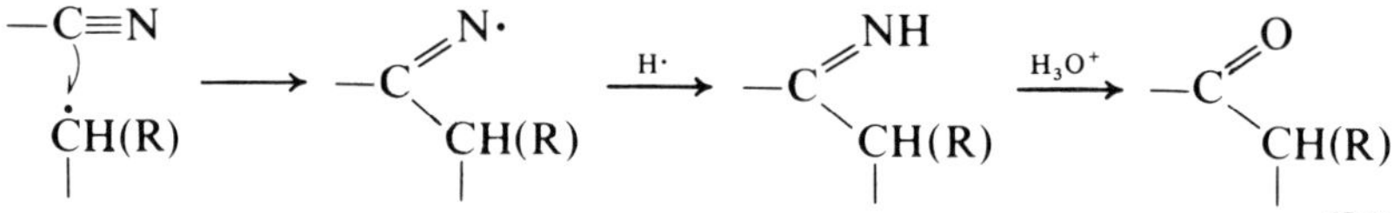

$$-C{\equiv}N \longrightarrow -C{=}N{\cdot} \xrightarrow{\text{H}\cdot} -C{=}NH \xrightarrow{\text{H}_3\text{O}^+} -C{=}O$$

(30)

$$R_F I + \longrightarrow R_F$$

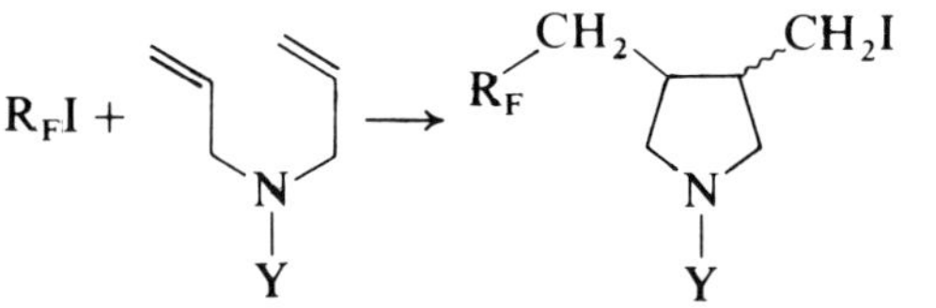

cis/trans

(31)

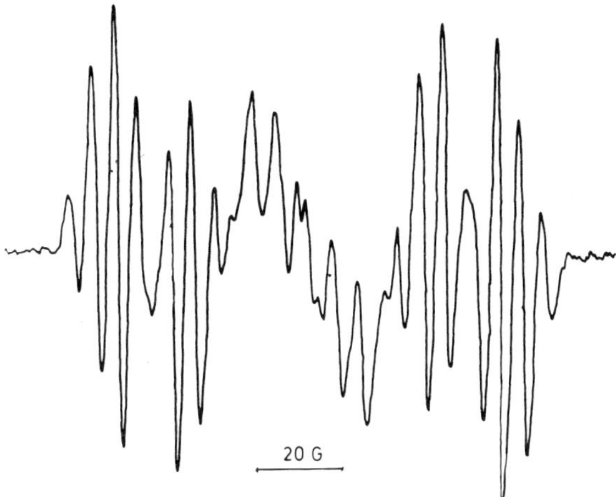

FIG. 3. Electron spin resonance spectrum of radicals from N,N-bis(2-isopropyl-allyl)methylamine and NH_3^+. (Reproduced from Beckwith, A. L. J., Hawthorne, D. G, and Solomon, D. H., *Aust. J. Chem.*, **29** (1976), p. 999, by permission of the publishers, CSIRO Ⓒ.)

LIX were found. The bicyclic adducts (**LV** and **LVI**) are the result of a *backbiting* reaction between the first-formed alkyl radical and the nitrile group, to form an imine which hydrolyses to a carbonyl function in one case (eqn. 30).

The following conclusions were drawn:

(i) in these systems, the initial attack occurs on a terminal carbon of the allyl groups;

(ii) bulky alkyl substituents on one allyl group result in a preference for the unsubstituted (least hindered) group;

(iii) intra-molecular cyclisation invariably leads to a five-membered ring structure;

(iv) where steric hindrance occurs, six-membered rings are found.

Brace has extended his earlier studies of the radical addition of perfluoroalkyl iodides to dienes to include a series of N-substituted diallylamines.[136] By spectroscopic and chemical methods, he has shown that five-membered ring adducts are formed (eqn. 31).

The stereochemistry of the product mixture was not discussed. However, in the reactions of diallyl ether with 1-iodo-F-butane, 1-iodo-F-hexane and 1-iodo-F-octane it is the *cis*-isomer which predominates, and pure samples of these isomers were isolated and characterised by ^{13}C n.m.r. This agrees with his earlier work on diallylcyanamide but conflicts with his data on 1,6-heptadiene where he reports that it is the *trans*-isomer which is formed most rapidly. Other workers[137] have observed that free-radical addition of 2-methylpropanoic acid to 1,6-heptadiene results in cyclisation to the five-membered ring adduct with a ratio of 2:1 in favour of the *trans*-isomer. As pointed out by Beckwith *et al.*,[121] the cyclisation of hex-5-enyl radicals is not always stereoselective towards the *cis*-isomer when the substituent at C-1 is bulky.

The products of free-radical cyclisation reactions (cyclotelomers) provide a valuable insight to the possible structures of analogous cyclopolymers. However, concern has been expressed[10,113,115] that cyclopolymerisation and cyclotelomerisation may not necessarily follow the same mechanism. For example, poly(divinylformal) yields cyclopolymers which are shown, by chemical methods, to contain up to 30 % five-membered ring units; yet in the cyclotelomers, five-membered rings predominate.[113] Clearly, direct analysis of cyclopolymers is the most desirable approach to determination of their structures.

3.2.3. Structures of Cyclopolymers and Cyclocopolymers

Nuclear magnetic resonance spectroscopy has proved invaluable for investigating in detail the structures of cyclopolymers and cyclocopolymers. Corfield and Monks[37] used ^{31}P n.m.r. spectroscopy to distinguish and quantitatively determine phosphorus atoms in five- and six-membered ring environments in the cyclopolymers from divinyl phosphonates (eqn. 32). Using a series of phosphonates as model compounds, it

$$\text{(32)}$$

R = Ph, CH$_3$

was shown by direct analysis of the polymers that poly(divinyl phenylphosphonate) contained five- and six-membered ring repeating units in amounts dependent upon the method of polymerisation, whereas poly(divinyl methylphosphonate) only contained six-membered rings.

Pulsed Fourier transform ^{13}C n.m.r. spectroscopy has been used most effectively to determine the structures of polymers of diallyl-amines.[12,15-17,135,138-140] The proton-decoupled ^{13}C n.m.r. spectrum of

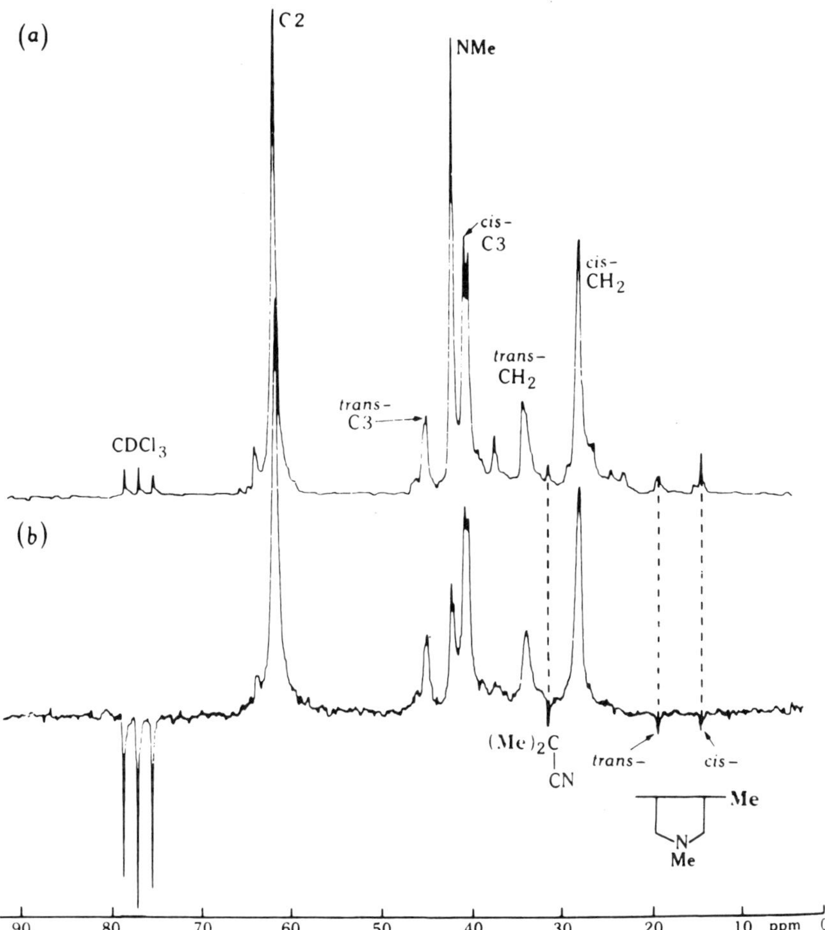

FIG. 4. Proton-decoupled ^{13}C n.m.r. spectra of a cyclopolymer of *N,N*-diallylmethylamine after a π–τ–$\pi/2$ sequence: (*a*) τ 20 s; (*b*) τ 0·2 s. (Reproduced from Hawthorne, D. G., Johns, S. R., Solomon, D. H. and Willing, R. I., *Aust. J. Chem.*, **32** (1979), p. 1156, by permission of the publishers, CSIRO ©.)

a cyclopolymer from *N,N*-diallylmethylamine is shown in Fig. 4(*a*). A comparison of this spectrum with those of a series of pyrrolidine and piperidine model compounds showed that the cyclic units in this polymer are five-membered rings (eqn. 33). The existence of both *cis*- and *trans*-units (ratio 5:1) was revealed from the chemical shifts of the methine carbons at positions 3,4 (*cis*, 41·7 ppm and *trans*, 45·7 ppm) and from the methylene carbons bridging the rings (*cis*, 28·4 ppm and *trans*, 34·4 ppm). Using an inversion/recovery technique[139] the signal at 31·7 ppm was assigned to the methyl groups of the initiating cyanopropyl radical and the signals at 14·6 and 19·3 ppm to *cis*- and *trans*-methyl groups in a terminal cyclic unit (Fig. 4(*b*)). The spectra of other *N*-substituted diallylamines were interpreted similarly.[17–19,21,138,140] The cyclisation reaction is influenced by substituents at the *β*-position of the allyl groups[15,16] (Table 3) and can yield five- or six-membered rings (eqn. 34) according to the substitution pattern.

$$(33)$$

$$(34)$$

LX

In the absence of appreciable steric hindrance, intra-molecular cyclisation proceeds to the five-membered ring. Where steric interactions are significant, six-membered rings are also observed. Changes in reaction temperature did not affect the structure of the cyclopolymers of *N,N*-diallylmethylamine. However, with *N,N*-dimethallylmethylamine changes in temperature did affect the ratio of five- to six-membered ring structures. The six-membered ring content increased with an increase in temperature as might be expected if the reaction becomes more subject to thermo-dynamic control (Table 4).

Tsukino and Kunitake[34] have investigated the microstructure of the cyclopolymers from divinyl formal, acetaldehyde divinyl acetal and acetone divinyl acetal (**LXI**, (*i*)–(*iii*)) using ^{13}C n.m.r. spectroscopy. By using a variety of model compounds, and from these calculating the chemical shifts for carbon atoms in a range of possible structures for the cyclopolymers, these authors determined that the polymers contained two types of structural unit. All the polymers contained the *cis*-4,5-disubstituted-1,3-dioxolane ring as the predominant unit in the main chain

TABLE 3

STRUCTURE OF CYCLOPOLYMERS FROM N,N-DIALLYLMETHYLAMINES
(**LX**)

Reactant (**LX**)		Cyclopolymer	
R_1	R_2	5-Ring (%)	6-Ring (%)
CH_3	H	100	—
CH_3	CH_3	50	50
CH_3CH_2	H	100	—
CH_3CH_2	CH_3CH_2	40	60
$(CH_3)_2CH$	H	100	—
$(CH_3)_3C$	H	50	50
C_6H_5	H	100	—
C_6H_5	C_6H_5	50	50
$CO_2CH_2CH_3$	H	100	—
$CO_2CH_2CH_3$	$CO_2CH_2CH_3$	—	100

(**LXIII**, 75%), with this structure being found also as a pendant group (**LXII**, 25%) formed by an intra-molecular chain-transfer mechanism, (eqn. 35). In poly(acetaldehyde divinyl acetal) the 2-methyl group may be *syn* or *anti*; the observed chemical shifts agree remarkably well with those estimated for the *cis–syn* form. The spectrum obtained for the cyclopolymer from divinyl formal is shown in Fig. 5, and the assignment of peaks is given in **LXII** and **LXIII**. These results differ from those obtained by chemical methods on similar polymers but this point is not taken up.

These authors suggest that similar structures to **LXII** are to be found in the cyclopolymers from N,N-diallylmethylamine and that the small peaks at 14·6 and 19·3 ppm in Fig. 4 should be assigned to methyl groups in a

TABLE 4

EFFECT OF TEMPERATURE ON STRUCTURE OF CYCLO-
POLYMERS FROM N,N-DIMETHALLYLMETHYLAMINE
(**LX**; $R_1 = R_2 = CH_3$)

Temperature (°C)	Cyclopolymer	
	5-Ring (%)	6-Ring (%)
30	60	40
60	50	50
80	50	50
130	40	60

pendant pyrrolidine ring (**LXIV**). However, the peaks in Fig. 4 are very much smaller than that at 14·0 ppm in Fig. 5, and a *trans*-methyl group is postulated for the peak at 19·3 ppm in Fig. 4, which is not as readily envisaged as resulting from intra-molecular chain transfer as a *cis*-methyl group. Since only one sample of each polymer has been investigated and correlation of these data with changes in molecular weight has not been studied, this problem requires further consideration.

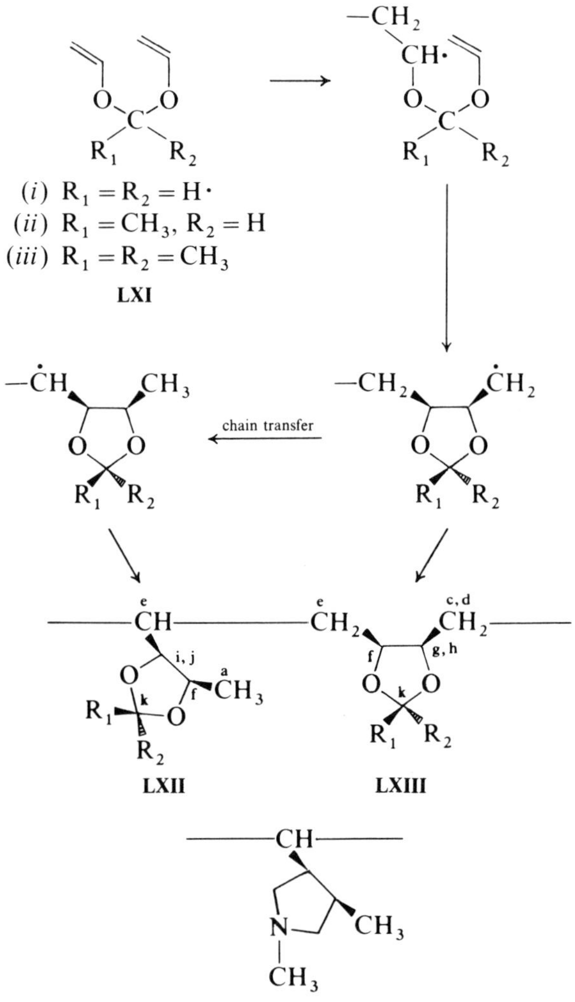

(*i*) $R_1 = R_2 = H \cdot$
(*ii*) $R_1 = CH_3, R_2 = H$
(*iii*) $R_1 = R_2 = CH_3$

LXI

chain transfer

(35)

LXII

LXIII

LXIV

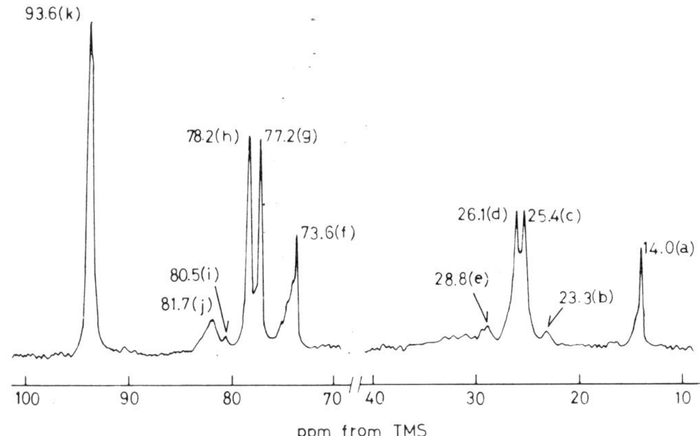

FIG. 5.   Carbon-13 n.m.r. spectrum of poly(divinyl formal). (Reproduced from Tsukino, M. and Kunitake, T., *Polymer J.*, 11 (1979), p. 440, by permission of the publishers, Society of Polymer Science, Japan ©.)

Tsukino and Kunitake[72] have also studied a cyclopolymer of divinyl ether using ^{13}C n.m.r. spectroscopy. Again the carbon chemical shifts for all the possible structural units and their stereoisomers were estimated, using model compounds as a basis for these calculations. Their conclusion was that only one kind of monocyclic intermediate is formed, and this is a five-membered ring product with *trans*-stereochemistry at the new bond (**LXVI** or **LXVII**, eqn. 36). Here again, there is a bulky substituent at the radical site and the cyclisation of radical (**LXV**) is not stereoselective towards *cis*-stereochemistry. The monocyclic units react with further monomer or cyclise to form bicyclic units (**LXVIII** or **LXIX**) in which the exocyclic methylene radical is in an *anti* position with respect to the preceding unit. The monocyclic and bicylic units are present in the polymer in approximately equal amounts.

The cyclocopolymer of divinyl ether and maleic anhydride has been studied by a number of authors in an attempt to clarify its structure. Samuels[141] has used the completely esterified copolymer as a molecular probe and carried out solution light scattering, gel permeation chromatography and intrinsic viscosity measurements. His results have shown that this polymer has a random coil conformation in solution and that there is the possibility that long-chain branching exists in higher molecular weight fractions. Butler[90] has suggested that pendant vinyl ether groups in units formed by the homopolymerisation of divinyl ether could act as branch points since maleic anhydride copolymerises readily with vinyl ethers.

Molecular model studies by Samuels revealed that the bulky ester groups caused much more steric hindrance in the tetrahydropyran unit (**II**) than in the alternative tetrahydrofuran unit (**IV**). The latter structure was more in accord with his experimental data; in particular, an acceptable characteristic ratio (C_∞) could be calculated for structure (**IV**) but not for (**II**).

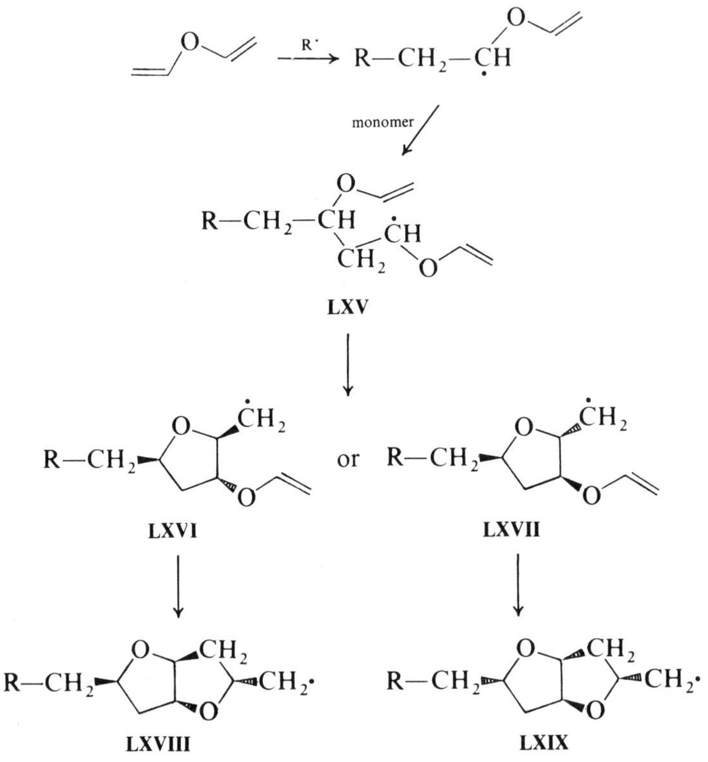

$$
\begin{array}{c}
\text{(reaction scheme)}
\end{array}
$$

LXV

LXVI or **LXVII**

LXVIII **LXIX**

(36)

Butler and Chu[142] have reported their results of a study of this cyclocopolymer. Using deuterated monomers, 300 MHz ^{1}H n.m.r. spectroscopy and ^{13}C n.m.r. spectroscopy, they concluded that a polymer prepared in cyclohexanone solution contained structure **II** (six-membered ring) with a *trans*-fused bicyclic unit (**LXX**) predominating over a *cis*-fused unit (**LXXI**), but that both *cis* and *trans* chain anhydride units were present, in approximately equal amounts, and there was no significant change in the structures observed for polymers produced at different temperatures.

Kunitake and Tsukino[143] have obtained differing results from a

[13]C n.m.r. study of this system. The carbon chemical shifts for possible units were calculated with the aid of model compounds. The major peaks of a cyclopolymer prepared in chloroform were consistent with the presence of a symmetrical bicyclic unit having *cis*-fused rings attached to a *trans* monocyclic anhydride (**LXXII**). A polymer produced in acetone/carbon disulphide was estimated to contain 90 % structure **II** (six-membered ring) and 10 % structure **IV** (five-membered ring).

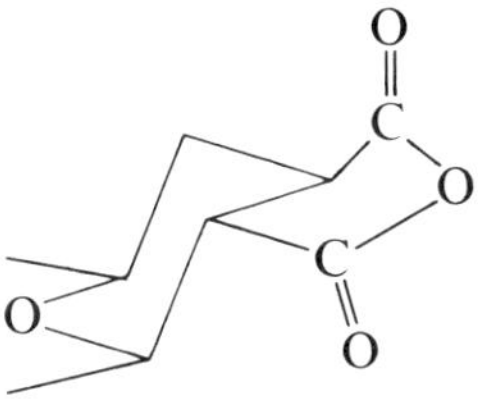

LXX

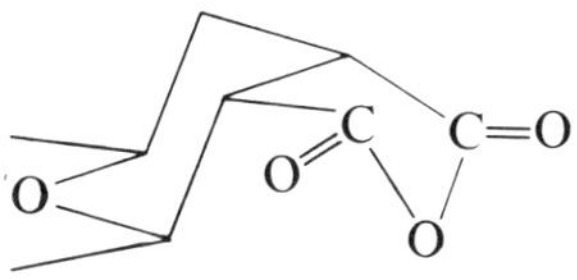

LXXI

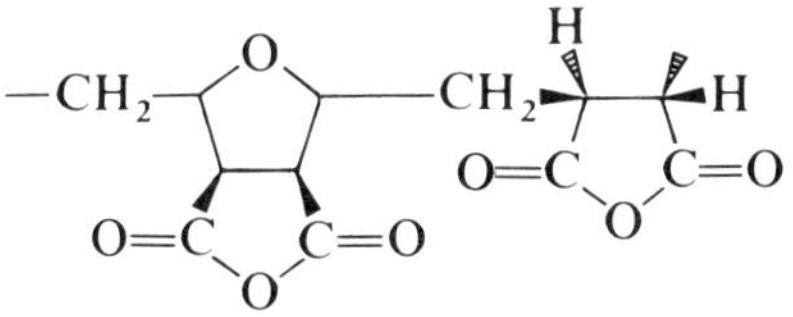

LXXII

Recently, Freeman and Breslow[144] have presented evidence, based primarily on [13]C n.m.r. data, that for a copolymer prepared in toluene or benzene, both structure **II** (six-membered ring) and structure **IV** (five-membered ring) are present in a ratio of $1\cdot0:0\cdot8$. Using model compounds, they assigned all of the ^{13}C peaks in the spectrum of the copolymer and the only unique (that is, without overlapping) carbon resonance is that of the methylene carbon in the six-membered ring. All spectra examined by these authors (including those published by Kunitake[143]) contain this peak. Further, these authors feel that there is insufficient spectroscopic evidence available, so far, for stereochemical assignments to be made.

To summarise, it is now clear that the original, not unreasonable, proposal of a six-membered ring as the fundamental unit of cyclopolymers and cyclocopolymers is not totally correct. In the polymerisation of diallylamines, which have been thoroughly studied, there is agreement that both five- and six-membered ring structures exist in amounts which are dependent upon the structure of the monomer and the conditions of the polymerisation process. In other cases, for example the cyclopolymers of divinyl acetals or the cyclocopolymers of divinyl ether and maleic anhydride, different approaches, using different samples of the polymers, have yielded conflicting results. A more systematic study of such examples, by complementary methods, is necessary before relationships to monomer structure and polymerisation process, or inadequacies in techniques, can be assessed. Obviously, the structures of cyclopolymers and cyclocopolymers cannot be predicted with certainty, as yet, but must be accurately determined if an understanding of their properties and the mechanism of this interesting process is to be gained.

4. MECHANISTIC INTERPRETATIONS

4.1. Cyclopolymerisation

There are three features which characterise the cyclopolymerisation reaction of non-conjugated dienes:

(i)   the intra-molecular cyclisation step in 1,6-dienes yields, in many cases, five-membered ring products and not the thermodynamically more stable six-membered rings;

(ii)   the intra-molecular reaction leading to soluble, essentially saturated, linear polymers predominates over the inter-molecular cross-linking reaction;

(iii)   the reactivity of dienes which cyclopolymerise is greater than that of mono-unsaturated compounds of similar structure.

4.1.1. Kinetic versus Thermodynamic Control

It is now well established that the ring size formed in free radical cyclisation reactions of 5-hexenyl radicals is influenced according to whether the reaction is rate- or equilibrium-controlled.[119,120,124,145] For example, cyclisation of the 5-hexenyl radical is an irreversible process leading to methylcyclopentane[111,146] (eqn. 37). But radicals stabilised by cyano and ester groups (eqn. 38) are reversibly cyclised to both five- and six-membered rings with the proportion of six-membered rings increasing with

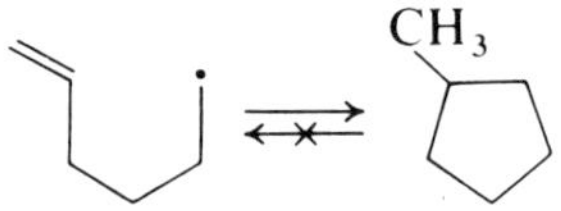

$$\text{(37)}$$

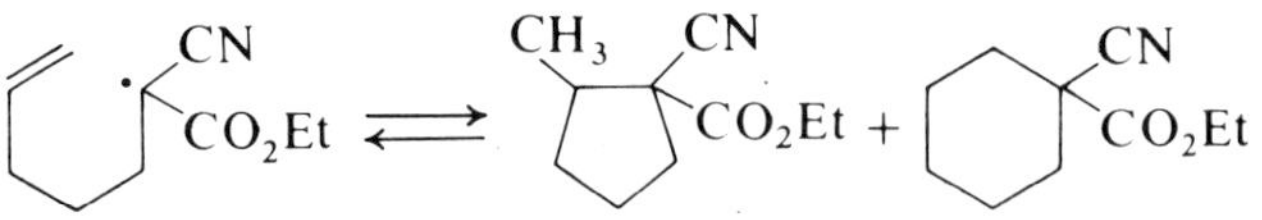

$$\text{(38)}$$

LXXIII   LXXIV   LXXV

TABLE 5

TEMPERATURE DEPENDENCE OF THE CYCLISATION OF **LXXIII**

Temperature (°C)	Products	
	5-Ring (**LXXIV**) (%)	6-Ring (**LXXV**) (%)
81	16	84
65	26	74
35	39	61
22	50	50
−70	80	20

an increase in temperature[119] (Table 5). Reversibility of this reaction was confirmed by decomposing the cyclic peresters **LXXVI** and **LXXVII** to discover that the product ratio was similar in each case (eqn. 39) and agreed closely with that from the open-chain radical (Table 5). With many alkenyl radicals the formation of a five-membered ring is kinetically favoured over

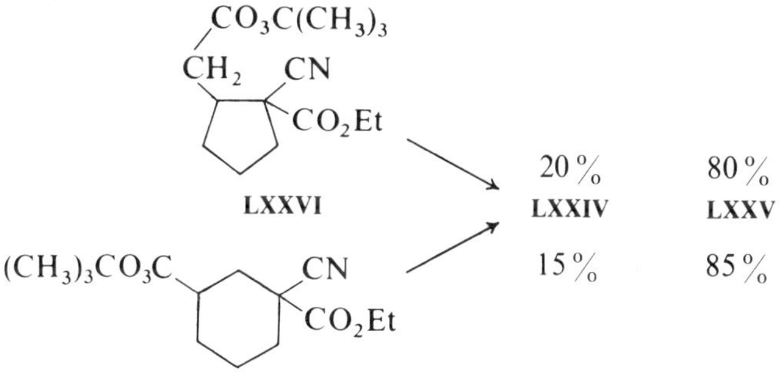

$$\begin{array}{cccc} & & 20\% & 80\% \\ \text{LXXVI} & & \text{LXXIV} & \text{LXXV} \end{array} \quad (39)$$

$$\begin{array}{cc} 15\% & 85\% \end{array}$$

LXXVII

formation of a six-membered ring.[121] However, with 5-substituted 5-hexenyl radicals the formation of a six-membered ring can be increased due to either the steric effect of the substituent causing a retardation in the rate of 1,5-cyclisation, as with **LXXVIII**,[123] or a resonance stabilising effect which favours 1,6-cyclisation, as with **LXXIX**.[128] 1,5-Ring closures of

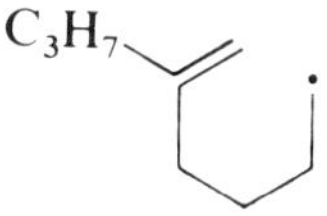

LXXVIII

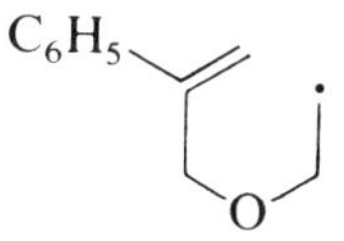

LXXIX

substituted 5-hexenyl and related radicals are stereoselective: 1- or 3-substituted systems afford mainly *cis*-disubstituted products, whereas 2- or 4-substituted systems give mainly *trans*-products.[121]

Studies of the cyclopolymerisation of variously substituted diallylamines suggest that these monomers also exhibit the characteristics of kinetic versus thermodynamic control of a reaction.[6,15–19,21,134,135,138,139,147] Consistent with kinetic control of the reaction, cyclopolymerisation of *N,N*-diallylmethylamine gave five-membered rings with a *cis*-configuration. Steric or conjugative factors, introduced by 2-substitution of one or both allyl groups, influence the initial attack on the monomer and the cyclisation reaction, where, with bulky or resonance stabilising groups at the point of attack, six-membered rings predominate (Table 3). The six-membered ring content of cyclopolymers from *N,N*-dimethallylmethylamine increases with an increase in the temperature of the reaction, as the higher-energy pathway becomes possible (Table 4).

It has been proposed that the kinetic preference for the formation of five-membered ring products is due to the stereo-electronic requirements of the transition state for radical addition reactions. Beckwith explains[121,124,129,131,148,149] that this concept demands maximum overlap of the half-filled *p* orbital (radical) with the vacant π^* (antibonding) orbital of the double bond. Approach of the radical must be along a vertical line from one of the carbon atoms of the double bond with the orbitals holding the three electrons being in the same plane throughout the reaction. These conditions can be readily fulfilled by 1,5-cyclisation of 5-hexenyl radicals but not with 1,6-cyclisation. It is suggested that the transition state is reactant-like with the 5-hexenyl system adopting a chair-like conformation (eqn. 40). With a substituent at position 1 (for example CH_3), it might be expected that the transition state for *trans*-addition (**LXXX**) would be more favourable, on steric grounds, than that for *cis*-addition (**LXXXI**). But *cis*-addition predominates in many radical cyclisation and cyclopolymerisation reactions. Beckwith has suggested[129] that further hyperconjugative

$$\tag{40}$$

mixing of the *p* orbital with the alkyl CH σ and σ^* orbitals (**LXXXII**), leading to a delocalised orbital of similar symmetry to the π^* orbital, results in an attractive interaction between the alkyl substituent and the double bond. However, this must not apply in cases where the 1-substituent is bulky or can stabilise the free electron by delocalisation, when *trans*-addition is preferred.[72,136,137,150,151]

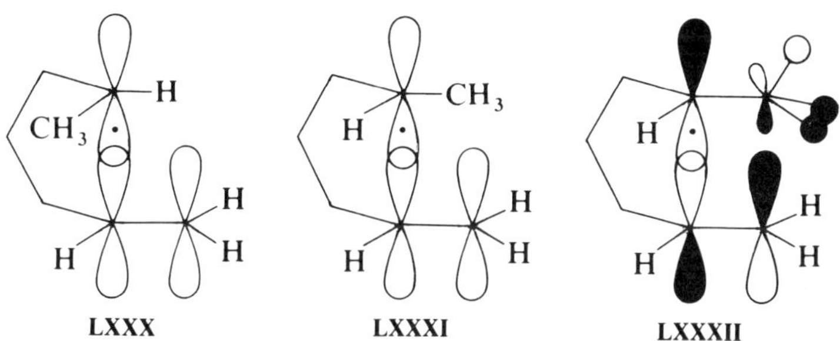

LXXX   **LXXXI**   **LXXXII**

This model for the cyclisation reaction can explain the increasing preference for cyclopolymers with six-membered rings as bulky 2-substituents are introduced to one or both allyl groups of diallylamines (Table 3). In the transition state to a five-membered ring there will be considerable interaction between the eclipsed groups (**LXXXIII**) which will be reduced in the transition state to a six-membered ring where the substituents become more staggered, as the system adopts a half-chair conformation (**LXXXIV**).

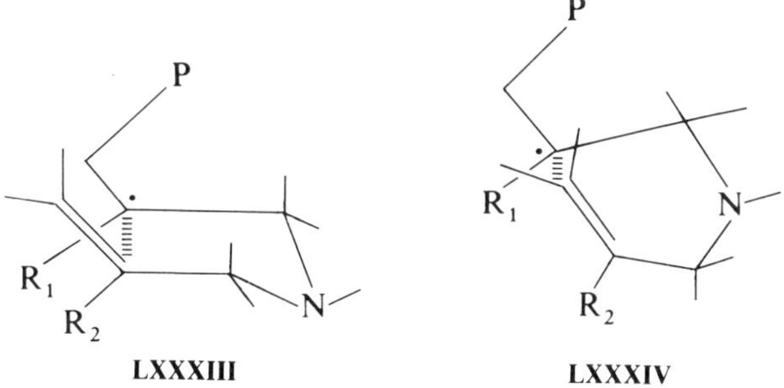

LXXXIII   **LXXXIV**

Other workers have provided support for this hypothesis to explain the cyclisation mechanism by examinations of molecular models,[145,152] calculations of geometrical probability factors[153] and molecular orbital evaluations.[125,140]

Although the ring size in cyclopolymers will be influenced by the nature of the cyclisation reaction, and it is therefore important to understand the energetics and kinetics of model reactions, there are indications that the correlation between models and polymers is unsatisfactory.[6,10,113,115,147] For example, the cyclisation of N,N-di(2-ethylallyl)methylamine produced a six-membered ring low molecular weight product, exclusively, but the cyclopolymer contained 40 % of a five-membered ring structure.[16] Perhaps this reflects the relative abilities of the intermediate radicals to undergo propagation and/or termination reactions. Alternatively, the factors controlling the cyclisation may be influenced by the growing chain, such that polymer microstructure is related to molecular weight. These points require investigation.

4.1.2. Electronic Interaction Theory

As a fundamental explanation for cyclopolymerisation, Butler[8] proposed that an electronic interaction occurs between the non-conjugated double bonds of 1,6-dienes, or between the intra-molecular double bond and the reactive centre after initiation, such interactions providing an energetically favourable pathway from monomer to cyclic product. This theory has been investigated many times, with conflicting results, and the earlier studies have been thoroughly reviewed.[4,7]

Further spectroscopic studies[154–156] have failed to produce evidence for interspatial interactions in monomers which cyclopolymerise. Kinetic evidence has been obtained[67,130] which shows that the reactivity of each double bond in the radical **LXXXV** is about five times greater than the double bond in the 5-hexenyl radical under similar conditions. At first this was attributed to a homoconjugative interaction which stabilised conformations favourable for intra-molecular reaction, and lowered the energy of the orbital involved in formation of the new bond. However, it was found that the 3-propyl-5-hexenyl radical also cyclised more rapidly than the 5-hexenyl radical and thus the enhanced reactivity of **LXXXV** cannot be attributed solely to homoconjugation.

1,3-Bis(4-vinylphenyl)propane can be polymerised, using cationic initiators, to a cyclopolymer having [3.3]paracyclophane units in the chain.[42] Cyclopolymers are not produced using radical or anionic initiators. As an explanation, an interaction is proposed, between the styryl type cation and the intra-molecular styryl group (**LXXXVI**), which stabilises the transition state leading to cyclopolymer.

Yokota *et al.*[56–60] have employed the technique of complexation with alkylaluminium chlorides, which is used for the preparation of alternating copolymers, to cyclopolymerise a number of unsymmetrical monomers. The ability of *o*-allylphenyl acrylate, 2-(*o*-allylphenoxy)ethyl acrylate, 4-(*o*-allylphenoxy)butyl acrylate and *o*-vinylphenyl acrylate to cyclopolymerise was increased when alkylaluminium chlorides were present. These results provide evidence for an intra-molecular interaction between the double bonds of these monomers, one of which is strongly electron accepting due to the complexing agent, and the other electron donating.

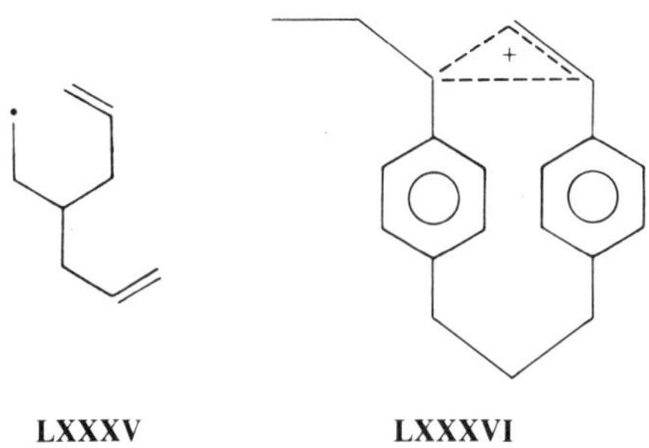

LXXXV　　　　　　　LXXXVI

If electronic interactions were the only explanation for cyclisation this would suggest a low-energy pathway to cyclopolymer. A study of the cyclopolymerisation of methacrylic anhydride[157] has shown that the intra-molecular step has a higher energy of activation ($10{\cdot}9\,kJ\,mol^{-1}$) than the inter-molecular step. Guaita *et al.* have determined similar values for acrylic anhydride ($9{\cdot}41\,kJ\,mol^{-1}$),[158] methacrylic anhydride ($9{\cdot}45\,kJ\,mol^{-1}$)[159] and *o*-divinylbenzene ($10{\cdot}29\,kJ\,mol^{-1}$).[41]

4.1.3. Steric Effects

A number of studies point towards a thermodynamic explanation for cyclopolymerisation. Although the differences in activation energies between intra-molecular and inter-molecular propagation reactions suggest that the former is a slightly less favourable reaction,[41,157–159] Guaita *et al.*[41,158,159] have explained the high tendency for cyclisation as due to a smaller decrease in entropy for such a reaction compared with inter-molecular addition. Thus, the Gibbs free energy of activation will be smaller for cyclopolymerisation than for inter-molecular propagation.

Kodaira has extended this hypothesis to propose that highly cyclised polymers may be obtained from monomers whose monofunctional counterparts do not homopolymerise. This has been successfully applied to a number of systems, including N-substituted dimethacrylamides,[30–32] N-allyl-N-methylmethacrylamide[63,64] and *sym*-dimethacryloyldimethyl-hydrazine.[160] Thus, N-isobutyryl-N-propylmethacrylamide did not homo-polymerise, but did copolymerise with styrene and methyl methacrylate, presumably for steric reasons. However, complete cyclisation occurs on homopolymerisation or copolymerisation of N-propyldimethacrylamide. The loss of internal rotation degrees of freedom is less effective in decreasing entropy than a loss of translational and rotational degrees of freedom, and the activation entropies favour cyclisation.

4.2. Cyclocopolymerisation

Mechanistic aspects of the cyclocopolymerisation reaction have been reviewed on a number of occasions.[4,90,161] The unusual nature of this reaction, in which bimolecular ring closure is highly favoured, led Butler to propose[11] that the propagating species is a charge-transfer complex. This proposal was supported by evidence for the existence of complexes in several comonomer pairs. Butler and Fujimori[162] have studied solvent effects on the cyclocopolymerisation of maleic anhydride and divinyl ether, and produced further evidence to support a charge-transfer complex as the active species in the polymerisation process. However, Zeegers and Butler[163] have concluded from kinetic data that the mechanism could be explained without involving a charge-transfer complex.

Although acrylonitrile forms a cyclocopolymer with divinyl ether, the structure is irregular due to the tendency of acrylonitrile to homo-polymerise. Addition of Lewis acids, such as zinc chloride and triethylaluminium, increased the rate of polymerisation and produced alternating copolymers (2:1) of acrylonitrile and divinyl ether. A charge-transfer complex was identified between divinyl ether and a complex of acrylonitrile with zinc chloride, which adds support to a mechanism involving a charge-transfer complex. In a detailed study of charge-transfer complex formation and cyclocopolymerisation between divinyl ether and several substituted maleic anhydrides, Fujimori and Butler[89] came to the following conclusions:

(i)   a strong charge-transfer complex produces a 1:1 cyclocopolymer;
(ii)   a sterically hindered anhydride yields a 1:1 cyclocopolymer;
(iii) a weak charge-transfer complex and a reactive anhydride yields a 1:2 cyclocopolymer.

Anhydrides, which fall into categories (i), (ii) and (iii) respectively, include chloromaleic anhydride, dimethylmaleic anhydride and maleic anhydride.

Fujimori[91] has detected charge-transfer complexes between divinyl sulphone and vinyl ethers and suggested that they are involved in the cyclocopolymerisation mechanism.

4.3. Conclusions

It is now clear that the structures of cyclopolymers and cyclocopolymers are dependent on the nature of the monomers and the reaction conditions. Predictions of structures based upon thermodynamic control of the cyclisation reaction can be misleading since kinetic products are often found. As a driving force for the cyclisation reaction, an intra-molecular interaction may be important, in some cases, particularly where a charge-transfer interaction can occur or is induced. However, the predominance of cyclisation over inter-molecular propagation can be ascribed to a smaller decrease in activation entropy which compensates for the unfavourable difference in activation energies. Further studies on the kinetics and energetics of the cyclisation reaction and the reactions of the competitive cyclic products, during polymerisation, are desirable.

Even 25 years after the early studies in this field, a considerable number of research papers and patents are published annually. The scope of the cyclopolymerisation and cyclocopolymerisation mechanisms for the synthesis of polymers with desirable structures and properties, the microstructure of the polymers and the driving force for these reactions are areas which are still actively investigated.

REFERENCES

1. BUTLER, G. B. and ANGELO, R. J., *J. Amer. Chem. Soc.*, **79** (1957), p. 3128.
2. BUTLER, G. B., CRAWSHAW, A. and MILLER, W. L., *J. Amer. Chem. Soc.*, **80** (1958), p. 3615.
3. BUTLER, G. B., *J. Macromol. Sci., Chem.*, **A5** (1971), p. 219.
4. BUTLER, G. B., CORFIELD, G. C. and ASO, C., *Progress in Polymer Science*, Volume 4, Jenkins, A. D. (Ed.), London, Pergamon Press, 1975, p. 71.
5. SOLOMON, D. H., *J. Polym. Sci., Polym. Symp.*, **49** (1975), p. 175; *J. Macromol. Sci., Chem.*, **A9** (1975), p. 97.
6. SOLOMON, D. H. and HAWTHORNE, D. G., *J. Macromol. Sci., Rev. Macromol. Chem.*, **C15** (1976), p. 143.
7. BUTLER, G. B., *Proc. Int. Symp. Macromol. 1974*, Mano, E. B. (Ed.), Amsterdam, Elsevier, 1975, p. 57; *J. Polym. Sci., Polym. Symp.*, **64** (1978), p. 71.

8. BUTLER, G. B., *J. Polym. Sci.*, **48** (1960), p. 279.
9. GIBBS, W. E. and BARTON, J. M., *Kinetics and Mechanism of Polymerization*, Volume 1, Part 1, Ham, G. E. (Ed.), New York, Dekker, 1967, Chapter 2.
10. CORFIELD, G. C., *Chem. Soc. Rev.*, **1** (1972), p. 523.
11. BUTLER, G. B., *Pure Appl. Chem.*, **23** (1970), p. 255.
12. OTTENBRITE, R. M. and RYAN, W. S., *Ind. Eng. Chem. Prod. Res. Dev.*, **19** (1980), p. 528.
13. BUTLER, G. B., (*IUPAC*) *Polymeric Amines and Ammonium Salts*, Goethals, E. J. (Ed.), London, Pergamon Press, 1980, p. 125.
14. SOLOMON, D. H., *J. Macromol. Sci., Chem.*, **A9** (1975), p. 95; *J. Macromol. Sci., Chem.*, **A10** (1976), p. 855.
15. HAWTHORNE, D. G. and SOLOMON, D. H., *J. Macromol. Sci., Chem.*, **A9** (1975), p. 149; *J. Macromol. Sci., Chem.*, **A10** (1976), p. 923.
16. HAWTHORNE, D. G., JOHNS, S. R., SOLOMON, D. H. and WILLING, R. I., *J. Chem. Soc., Chem. Commun.*, 1975, p. 982; *Aust. J. Chem.*, **29** (1976), p. 1955.
17. JOHNS, S. R., WILLING, R. I., MIDDLETON, S. and ONG, A. K., *J. Macromol. Sci., Chem.*, **A10** (1976), p. 875.
18. HODGKIN, J. H. and SOLOMON, D. H., *J. Macromol. Sci., Chem.*, **A10** (1976), p. 893.
19. HODGKIN, J. H. and ALLAN, R. J., *J. Macromol. Sci., Chem.*, **A11** (1977), p. 937.
20. HODGKIN, J. H., *Polym. Preprints., Amer. Chem. Soc., Div. Polym. Chem.*, **19** (1978), p. 420; *Chem. Ind. (London)*, 1979, p. 153.
21. HODGKIN, J. H. and DEMERAC, S., *Amer. Chem. Soc., Adv. Chem. Ser.*, **187** (1980), p. 211.
22. McLEAN, C. D., ONG, A. K. and SOLOMON, D. H., *J. Macromol. Sci., Chem.*, **A10** (1976), p. 857.
23. JACKSON, M. B., *J. Macromol. Sci., Chem.*, **A10** (1976), p. 959.
24. JACKSON, M. B., *J. Macromol. Sci., Chem.*, **A12** (1978), p. 853.
25. EPPINGER, K. H. and JACKSON, M. B., *J. Macromol., Sci., Chem.*, **A14** (1980), p. 121.
26. WYROBA, A., *Polimery (Warsaw)*, **23** (1978), p. 86.
27. JAEGER, W., WANDREY, C., REINISCH, G. and LINOW, K. J., *Faserforsch. Textiltech.*, **29** (1978), p. 647.
28. TOPCHIEV, D. A., BIKASHEVA, G. T., MARTYNENKO, A. I., KAPTSOV, N. N., GUDKOVA, L. A. and KABANOV, V. A., *Vysokomol. Soedin., Ser. B.*, **22** (1980), p. 269.
29. DANIELYAN, V. A., KARSLYAN, S. V. and MATSOYAN, S. G., *Arm. Khim. Zh.*, **32** (1979), p. 970.
30. KODAIRA, T. and AOYAMA, F., *J. Polym. Sci., Polym. Chem.*, **12** (1974), p. 897.
31. KODAIRA, T., NI-IMOTO, M. and AOYAMA, F., *Makromol. Chem.*, **179** (1978), p. 1791.
32. KODAIRA, T. and SAKAI, M., *Polym. J.*, **11** (1979), p. 595.
33. YAMAKITA, H. and HAYAKAWA, K., *Nippon Kagaku Kaishi*, **5** (1977), p. 706.
34. TSUKINO, M. and KUNITAKE, T., *Polym. J.*, **11** (1979), p. 437.
35. KIDA, S., NOZAKURA, S. and MURAHASHI, S., *Polym. J.*, **3** (1972), p. 234.
36. BILLINGHAM, N. C., JENKINS, A. D., KRONFLI, E. B. and WALTON, D. R. M., *J. Polym. Sci., Polym. Chem.*, **15** (1975), p. 675.

37. CORFIELD, G. C. and MONKS, H. H., *J. Macromol. Sci., Chem.*, **A9** (1975), p. 1113.

38. KUNITAKE, T., NAKASHIMA, T. and ASO, C., *J. Polym. Sci., Part A-1*, **8** (1970), p. 2853; *Makromol. Chem.*, **146** (1971), p. 79; ASO, C., KUNITAKE, T. and TAGAMI, S., *Progress in Polymer Science, Japan*, Volume 1, IMOTO, M. and ONOGI, S. (Eds.), New York, Halsted Press, 1971, p. 170.

39. SOSIN, S. L., JASHI, L. V., ANTIPOVA, B. A. and KORSHAK, V. V., *Vysokomol. Soedin., Ser. B*, **12** (1970), p. 699; *Vysokomol. Soedin., Ser. B*, **16** (1974), p. 347; KORSHAK, V. V. and SOSIN, S. L., *Organometallic Polymers*, CARREHER, C. E., SHEATS, J. E. and PITTMAN, C. U. (Eds.), London, New York, Academic Press, 1977, p. 26.

40. CORFIELD, G. C., BROOKS, J. S. and PLIMLEY, S., *Polym. Preprints, Amer. Chem. Soc., Div. Polym. Chem.*, **22** (1981), p. 3.

41. COSTA, L., CHIANTORE, O. and GUAITA, M., *Polymer*, **19** (1978), p. 197, p. 202.

42. NISHIMURA, J. and YAMASHITA, S., *Polym. Preprints, Amer. Chem. Soc., Div. Polym. Chem.*, **22** (1981), p. 46.

43. NISHIKUBO, T., IIZAWA, T. and YOSHINAGA, A., *Makromol. Chem.*, **180** (1979), p. 2793.

44. SEUNG, S. L. N. and YOUNG, R. N., *J. Polym. Sci., Polym. Lett.*, **16** (1978), p. 367.

45. MATHIAS, L. J. and CANTERBERRY, J. B., *Polym. Preprints, Amer. Chem. Soc., Div. Polym. Chem.*, **22** (1981), p. 38.

46. CHU, S-C. and BUTLER, G. B., *J. Polym. Sci., Polym. Lett.*, **15** (1977), p. 277.

47. BUTLER, G. B. and LIEN, Q. S., *Polym. Preprints, Amer. Chem. Soc., Div. Polym. Chem.*, **22** (1981), p. 54.

48. GUENIFFEY, H., KAEMMERER, H. and PINAZZI, C., *Makromol. Chem.*, **165** (1973), p. 73; KAEMMERER, H., STEINER, V., GUENIFFEY, H. and PINAZZI, C., *Makromol. Chem.*, **177** (1977), p. 1665.

49. ASINOVSKAYA, D. N., ANDREEV, D. N. and DAVIDYUK, L. N., *Vysokomol. Soedin., Ser. A.*, **19** (1977), p. 1325.

50. NISHINO, J., YOSHIDA, T., MAKINO, I., NANYA, S., TAMAKI, K. and SAKAGUCHI, Y., *Kobunshi Ronbunshu*, **31** (1974), p. 177.

51. KIKUKAWA, K., NOZAKURA, S. and MURAHASHI, S., *J. Polym. Sci., Part A-1*, **10** (1972), p. 139.

52. MATSUMOTO, A., IWANAMI, K. and OIWA, M., *J. Polym. Sci., Polym. Lett.*, **18** (1980), p. 307.

53. MATSUMOTO, A., IWANAMI, K. and OIWA, M., *J. Polym. Sci., Polym. Chem.*, **19** (1981), p. 213.

54. MATSUMOTO, A., IWANAMI, K., KITAMURA, T., OIWA, M. and BUTLER, G. B., *Polym. Preprints, Amer. Chem. Soc., Div. Polym. Chem.*, **22** (1981), p. 36.

55. URUSHIDO, K., MATSUMOTO, A. and OIWA, M., *J. Polym. Sci., Polym. Lett.*, **19** (1981), p. 59.

56. YOKOTA, K. and TAKADA, Y., *Kobunshi Kagaku*, **30** (1973), p. 71; *Kobunshi Kagaku*, **30** (1973), p. 217.

57. YOKOTA, K., KANEKO, N. and TAKADA, Y., *Kobunshi Kagaku*, **30** (1973), p. 475.

58. YOKOTA, K., HIRAYAMA, N. and TAKADA, Y., *Polym. J.*, **7** (1975), p. 629.

59. YOKOTA, K., KAKUCHI, T. and TAKADA, Y., *Polym. J.*, **8** (1976), p. 495.

60. YOKOTA, K., KAKUCHI, T. and TAKADA, Y., *Polym. J.*, **10** (1978), p. 19.
61. KAKUCHI, T., YOKOTA, K. and TAKADA, Y., *Polym. J.*, **11** (1979), p. 7.
62. YOKOTA, K., KANEKO, N., IWATA, J., KOMURO, K. and TAKADA, Y., *Polym. J.*, **11** (1979), p. 929.
63. KODAIRA, T., ISHIKAWA, M. and MURATA, O., *J. Polym. Sci., Polym. Chem.*, **14** (1976), p. 1107.
64. KODAIRA, T. and MURATA, O., *J. Polym. Sci., Polym. Chem.*, **17** (1979), p. 319.
65. PANZIG, H. L. and MULVANEY, J. E., *J. Polym. Sci., Polym. Chem.*, **10** (1972), p. 3469.
66. BECKWITH, A. L. J., ONG, A. K. and SOLOMON, D. H., *J. Macromol. Sci., Chem.*, **A9** (1975), p. 125.
67. BECKWITH, A. L. J. and MOAD, G., *J. Chem. Soc., Perkin II* (1975), p. 1726.
68. ALIEVA, A. G., STOTSKAYA, L. L. and KRENTSEL, B. A., *Vysokomol. Soedin., Ser. A.*, **15** (1973), p. 1005.
69. HEUBLIN, G. and HENBLEIN, B., *Plaste Kautsch.*, **23** (1976), p. 95.
70. VAN HEININGEN, J. J. and BUTLER, G. B., *J. Macromol. Sci., Chem.*, **A8** (1974), p. 1175.
71. PINAZZI, C. P., CATTIAUX, J. and BROSSE, J. C., *Eur. Polym. J.*, **10** (1974), p. 837.
72. KUNITAKE, T. and TSUKINO, M., *Makromol. Chem.*, **177** (1976), p. 303; TSUKINO, M. and KUNITAKE, T., *Macromolecules*, **12** (1979), p. 387.
73. GUAITA, M., CAMINO, G., CHIANTORE, O., REVELLINO, M. and TROSSARELLI, L., *Makromol. Chem.*, **175** (1974), p. 457.
74. ZAIMA, T., NISHIKUBO, T. and MITSUHASHI, K., *J. Polym. Sci., Polym. Lett.*, **18** (1980), p. 1.
75. AKOPYAN, L. A., AMBARTSUMYAN, G. V., OVAKIMYAN, E. V. and MATSOYAN, S. G., *Vysokomol. Soedin., Ser. A.*, **19** (1977), p. 271.
76. AKOPYAN, L. A., AMBARTSUMYAN, G. V., GRIGORYAN, S. G. and MATSOYAN, S. G., *Vysokomol. Soedin., Ser. A.*, **19** (1977), p. 1068.
77. AKOPYAN, L. A., AMBARTSUMYAN, G. V., MATSOYAN, M. S., OVAKIMYAN, E. V. and MATSOYAN, S. G., *Arm. Khim. Zh.*, **30** (1971), p. 771.
78. AKOPYAN, L. A., AMBARTSUMYAN, G. V., OVAKIMYAN, E. V., GEVORKYAN, S. B. and MATSOYAN, S. G., *Arm. Khim. Zh.*, **31** (1978), p. 510.
79. GIBSON, H. W., BAILEY, F. C., EPSTEIN, A. J., ROMMELMANN, H. and POCHAN, J. M., *J. Chem. Soc., Chem. Commun.* (1980), p. 426.
80. GIBSON, H. W., BAILEY, F. C., POCHAN, J. M. and HARBOUR, J., *Polym. Preprints, Amer. Chem. Soc., Div. Polym. Chem.*, **22** (1981), p. 35.
81. ONOSOV, G. V., SOROKIN, M. F., SHODE, L. G. and DOBROVINSKII, L. A., *Tr. Mosk. Khim.-Tekhnol. Inst.*, **86** (1975), p. 106; SOROKIN, M. F., SHODE, L. G. and ONOSOV, G. V., *Deposited Doc.*, 1976, Viniti 2197-76.
82. TAGAMI, S. and KUNITAKE, T., *Makromol. Chem.*, **175** (1974), p. 3367.
83. BUR, A. J. and FETTERS, L. J., *Chem. Rev.*, **76** (1976), p. 727.
84. WOEHRLE, D., *Makromol. Chem.*, **161** (1972), p. 121.
85. SAYADYAN, A. G., DZHANIKYAN, O. A. and MIRZOYAN, V. A., *Arm. Khim. Zh.*, **29** (1976), p. 708; SAFARYAN, E. B. and SAYADYAN, A. G., *Arm. Khim. Zh.*, **32** (1979), p. 401; SAFARYAN, E. B. and SAYADYAN, A. G., *Arm. Khim. Zh.*, **32** (1979), p. 405.
86. AMEMIYA, Y., KATAYAMA, M. and HARADA, S., *Makromol. Chem.*, **176** (1975),

p. 1289; AMEMIYA, Y., KATAYAMA, M. and HARADA, S., *Makromol. Chem.*, **178** (1977), p. 289.

87. FUJIMORI, K. and BUTLER, G. B., *J. Macromol. Sci., Chem.*, **A6** (1972), p. 1609.
88. FUJIMORI, K. and BUTLER, G. B., *J. Macromol. Sci., Chem.*, **A7** (1973), p. 415.
89. FUJIMORI, K. and BUTLER, G. B., *J. Macromol. Sci., Chem.*, **A7** (1973), p. 387.
90. BUTLER, G. B., *Anionic Polymeric Drugs*, Donaruma, L. G., Ottenbrite, R. M. and Vogl, O. (Eds.), New York, Wiley–Interscience, 1980, p. 49.
91. FUJIMORI, K., *J. Macromol. Sci., Chem.*, **A10** (1976), p. 1005.
92. KUNITAKE, T., YAMAGUCHI, K. and ASO, C., *Makromol. Chem.*, **172** (1973), p. 85.
93. AMEMIYA, Y., KATAYAMA, M. and HARADA, S., *Makromol. Chem.*, **178** (1977), p. 2499.
94. BUTLER, G. B., US Patent 3 288 770, 29 November 1966.
95. HOOVER, M. F. and CARR, H. E., *J. Tech. Assoc. of Pulp and Paper Ind.*, **51** (1968), p. 556.
96. HARADA, S. and KATAYAMA, M., *Makromol. Chem.*, **90** (1966), p. 177.
97. BUTLER, G. B. and BUNCH, R. L., *J. Amer. Chem. Soc.*, **71** (1949), p. 3120.
98. CLINGMAN, A. L., PARRISH, J. R. and STEVENSON, R., *J. Appl. Chem.*, **13** (1963), p. 1.
99. BATTAERD, H. A. J., US Patent 3 619 394, 9 November 1971.
100. WEISS, D. E., BOLTO, B. A., MCNEILL, R. A., MACPHERSON, A. S., SIUDAK, R., SWINTON, E. A. and WILLIS, D., *Aust. J. Chem.*, **19** (1966), p. 561, p. 589, p. 765 and p. 791.
101. BOLTO, B. A., (*IUPAC*) *Polymeric Amines and Ammonium Salts*, Goethals, E. J. (Ed.), London, Pergamon Press, 1980, p. 365.
102. BRESLOW, D. S., *Pure Appl. Chem.*, **46** (1976), p. 103.
103. BRESLOW, D. S., *Polymeric Preprints, Amer. Chem. Soc., Div. Polym. Chem.*, **22** (1981), p. 24.
104. MARVEL, C. S. and VEST, R. D., *J. Amer. Chem. Soc.*, **79** (1957), p. 5771; MARVEL, C. S. and STILLE, J. K., *J. Amer. Chem. Soc.*, **80** (1958), p. 1740; MARVEL, C. S. and GALL, E. J., *J. Org. Chem.*, **25** (1960), p. 1784.
105. FRIEDLANDER, W. S., *Abstracts, 133rd American Chemical Society Meeting, San Francisco, 1958*, p. 18N; FRIEDLANDER, W. S. and TIERS, G. V. D., German Patent 1 098 942.
106. ARBUZOVA, I. A. and SULTANOV, K., *Vysokomol. Soedin.*, **2** (1960), p. 1077; SULTANOV, K. and ARBUZOVA, I. A., *Uzbek. Khim. Zhur.*, **7** (1963), p. 57.
107. MATSOYAN, S. G., *Russ. Chem. Rev.*, **35** (1966), p. 32.
108. MURAHASHI, S., NOZAKURA, S., FUJI, S. and KIKUKAWA, K., *Bull. Chem. Soc. Japan*, **38** (1965), p. 1905.
109. MERCIER, J. and SMETS, G., *J. Polym. Sci.*, **57** (1962), p. 763.
110. SOKOLOVA, T. A. and RUDKOVSKAYA, G. D., *J. Polym. Sci., Part C, Polymer Symp.*, **16** (1967), p. 1157.
111. LAMB, R. C., AYERS, P. W. and TONEY, M. K., *J. Amer. Chem. Soc.*, **85** (1963), p. 3483.
112. BRACE, N. O., *J. Amer. Chem. Soc.*, **86** (1964), p. 523; *J. Org. Chem.*, **31** (1966), p. 2879; *J. Org. Chem.*, **32** (1967), p. 2711; *J. Org. Chem.*, **34** (1969), p. 2441; *J. Polym. Sci., Part A-1*, **8** (1970), p. 2091.
113. ASO, C., *Pure Appl. Chem.*, **23** (1970), p. 287.

114. WALLING, C., COOLEY, J. H., PONARAS, A. A. and RACAH, E. J., *J. Amer. Chem. Soc.*, **88** (1966), p. 5361.

115. DE WITTE, E. and GOETHALS, E. J., *J. Macromol. Sci., Chem.*, **A5** (1971), p. 73.

116. DE WITTE, E. and GOETHALS, E. J., *Makromol. Chem.*, **115** (1968), p. 234.

117. BOOTH, H., BOSTOCK, A. H., FRANKLIN, N. C., GRIFFITHS, D. V. and LITTLE, J. H., *J. Chem. Soc., Perkin II* (1978), p. 899.

118. WALLING, C., *Molecular Rearrangements*, Volume 1, de Mayo, P. (Ed.), New York, Interscience, 1963, Chapter 7.

119. JULIA, M., *Pure Appl. Chem.*, **15** (1967), p. 167; *Acc. Chem. Res.*, **4** (1971); p. 386.

120. WILT, J. W., *Free Radicals*, Volume 1, Kochi, J. K. (Ed.), New York, Wiley–Interscience, 1973, Chapter 8.

121. BECKWITH, A. L. J., EASTON, C. J. and ALGIRDAS, K. S., *J. Chem. Soc., Chem. Commun.* (1980), p. 482.

122. BECKWITH, A. L. J. and MOAD, G., *J. Chem. Soc., Chem. Commun.* (1974), p. 472.

123. BECKWITH, A. L. J., BLAIR, I. A. and PHILLIPOU, G., *Tetrahedron Letters*, 1974, p. 2251.

124. BECKWITH, A. L. J., *Essays in Free-radical Chemistry*, Special Publication 24, London, The Chemical Society, 1970, p. 239.

125. DEWAR, M. J. S. and OLIVELLA, S., *J. Amer. Chem. Soc.*, **100** (1978), p. 5290.

126. JULIA, M., DESCOINS, C., BAILLARGE, M., JACQUET, B., UGVEN, D. and GROEGER, F. A., *Tetrahedron*, **31** (1975), p. 1737.

127. WALLING, C. and CIOFFARI, A., *J. Amer. Chem. Soc.*, **94** (1972), p. 6059.

128. SMITH, T. W. and BUTLER, G. B., *J. Org. Chem.*, **43** (1978), p. 6.

129. BECKWITH, A. L. J., BLAIR, I. and PHILLIPOU, G., *J. Amer. Chem. Soc.*, **96** (1974), p. 1613.

130. BECKWITH, A. L. J., LAWRENCE, T. and SERELIS, A. K., *J. Chem. Soc., Chem. Commun.* (1980), p. 484.

131. BECKWITH, A. L. J. and WAGNER, R. D., *J. Amer. Chem. Soc.*, **101** (1979), p. 7099; *J. Chem. Soc., Chem. Commun.* (1980), p. 485.

132. BECKWITH, A. L. J., ONG, A. K. and SOLOMON, D. H., *J. Macromol. Sci., Chem.*, **A9** (1975), p. 115.

133. BECKWITH, A. L. J., HAWTHORNE, D. G. and SOLOMON, D. H., *Aust. J. Chem.*, **29** (1976), p. 995.

134. JOHNS, S. R. and WILLING, R. I., *J. Macromol. Sci., Chem.*, **A9** (1975), p. 169.

135. HAWTHORNE, D. G., JOHNS, S. R. and WILLING, R. I., *Aust. J. Chem.*, **29** (1976), p. 315.

136. BRACE, N. O., *J. Org. Chem.*, **36** (1971), p. 3187; *J. Org. Chem.*, **44** (1979), p. 212; *J. Org. Chem.*, **38** (1973), p. 3167.

137. BRADNEY, M. A. M., FORBES, A. D. and WOOD, J., *J. Chem. Soc., Perkin II* (1973), p. 1655.

138. LANCASTER, J. E., BACCEI, L. and PANZER, H. P., *J. Polym. Sci., Polym. Lett.*, **14** (1976), p. 549.

139. HAWTHORNE, D. G., JOHNS, S. R., SOLOMON, D. H. and WILLING, R. I., *Aust. J. Chem.*, **32** (1979), p. 1155.

140. OTTENBRITE, R. M. and SHILLADY, D. D., *(IUPAC) Polymeric Amines and Ammonium Salts*, Goethals, E. J. (Ed.), London, Pergamon Press, 1980,

p. 143; OTTENBRITE, R. M., *Polym. Preprints, Amer. Chem. Soc., Div. Polym. Chem.*, **22** (1981), p. 42.

141. SAMUELS, R. J., *Polymer*, **18** (1977), p. 453.

142. BUTLER, G. B. and CHU, Y. C., *J. Polym. Sci., Polym. Chem.*, **17** (1979), p. 859.

143. KUNITAKE, T. and TSUKINO, M., *J. Polym. Sci., Polym. Chem.*, **17** (1979), p. 877.

144. FREEMAN, W. J. and BRESLOW, D. S., *Preprints, Amer. Chem. Soc., Div. Org. Coatings and Plastics Chem.*, **44** (1981), p. 108.

145. JULIA, M., *Pure Appl. Chem.*, **40** (1974), p. 553.

146. KOCHI, J. K. and KRUSIC, P. J., *J. Amer. Chem. Soc.*, **91** (1969), p. 3940.

147. HAWTHORNE, D. G. and SOLOMON, D. H., *J. Polym. Sci., Polym. Symp.*, **55** (1976), p. 211.

148. STRUBLE, D. L., BECKWITH, A. L. J. and GREAM, G. E., *Tetrahedron Lett.*, 1968, p. 3701.

149. BECKWITH, A. L. J., GREAM, G. E. and STRUBLE, D. S., *Aust. J. Chem.*, **25** (1972), p. 1081.

150. JULIA, M. and MAUMY, M., *Bull. Soc. Chim. Fr.*, 1966, p. 434.

151. WALLING, C. and CIOFFARI, A., *J. Amer. Chem. Soc.*, **94** (1972), p. 6064.

152. BUTLER, G. B. and RAYMOND, M. A., *J. Polym. Sci., Part A*, **3** (1965), p. 3413.

153. HAMAAN, S. D., POMPE, A., SOLOMON, D. H. and SPURLING, T. H., *Aust. J. Chem.*, **29** (1976), p. 1975.

154. BUTLER, G. B. and VAN HEININGEN, J. J., *J. Macromol. Sci., Chem.*, **A8** (1974), p. 1139.

155. BAUCOM, K. B. and BUTLER, G. B., *J. Macromol. Sci., Chem.*, **A8** (1974), p. 1205.

156. DONESCU, D., CARP, N. and GOSA, K., *Rev. Roumaine Chim.*, **24** (1979), p. 501.

157. GRAY, T. F. and BUTLER, G. B., *J. Macromol. Sci., Chem.*, **A9** (1975), p. 45.

158. GUAITA, M., *Makromol. Chem.*, **157** (1972), p. 111.

159. CHIANTORE, O., CAMINO, G., CHIORINO, A. and GUAITA, M., *Makromol. Chem.*, **178** (1977), p. 125.

160. KODAIRA, T., SAKAI, M. and YAMAZAKI, K., *J. Polym. Sci., Polym. Lett.*, **13** (1975), p. 521; KODAIRA, T., YAMAZAKI, K. and KITOH, T., *Polym. J.*, **11** (1979), p. 377.

161. BUTLER, G. B., *Polyelectrolytes and Their Applications*, Rembaum, A. and Seligny, E. (Eds.), Dordrecht, Reidel Publishing Co., 1975, p. 97.

162. BUTLER, G. B. and FUJIMORI, K., *J. Macromol. Sci., Chem.*, **A6** (1972), p. 1553.

163. ZEEGERS, B. and BUTLER, G. B., *J. Macromol. Sci., Chem.*, **A6** (1972), p. 1569.

Chapter 2

PHOTOCHEMICAL CROSS-LINKING IN POLYMER-BASED SYSTEMS

A. Ledwith

University of Liverpool, UK

SUMMARY

The general nature and properties of cross-linked polymers, together with the most important factors governing light absorption and dissipation, are outlined in a brief introduction in order to illustrate how light may be used to bring about cross-linking. Two major types of photo-cross-linking involve step-growth and chain-growth processes respectively and these are described with examples of appropriate functional groups. Important applications of photochemically active polymer systems, including photocure of surface coatings and photoresist technology, are noted.

1. GENERAL INTRODUCTION

The oldest recorded example of photochemically induced cross-linking in a polymer-based system goes back to pre-Biblical times when, for the preparation of Egyptian mummies, it is recorded that linen cloths were dipped into a solution of lavender oil containing Syrian asphalt. The latter is a bituminous material, which hardens under the influence of sunlight, and is known to contain a variety of unsaturated aromatic, heterocyclic and other compounds. A similar photohardening of asphalt was used in the first imaging material for permanent photographs by Niepce as early as 1822. Later on in the nineteenth century, a second photo-cross-linking system was discovered involving immobilisation of gelatin when exposed to sunlight in the presence of potassium dichromate. It is now known that gelatin may be replaced by poly(vinyl alcohol) or other natural and synthetic polymers with hydroxylic, amino, amide and/or carboxyl groups. Although the detailed chemistry of these systems is still not fully understood it is clear that photolysis consistently reduced the dichromate ion and oxidised the polymer. This type of photo-cross-linking formed the basis of the earliest types of photoresist (see later) and is still used in commercial systems.[1]

2.   THE NATURE AND PROPERTIES OF CROSS-LINKED POLYMERS

Macromolecules are formed by two main processes usually classified (for kinetic reasons) into step-reaction polymerisations and chain-reaction polymerisations. In the former, the polymer chain grows from monomer molecules by a succession of repetitive reactions yielding stable molecules after each step in which the degree and type of functionality remains constant. In contrast, chain-reaction polymerisations of olefins and strained cyclic compounds must be catalysed or initiated by a reactive species independent of the monomer molecules. The polymer chains grow by rapid addition of monomer molecules to a propagating chain end carrying the same type of activity originally present in the catalyst or initiator fragment. Activity remains at the growing chain end of the polymer molecule until it is terminated or transformed by a variety of chemical reactions with components of the system. The formation of networks in the absence of photo-initiation is described for stepwise reactions in Chapter 3 and for chain reactions in Chapter 4.

Thus, the formation of cross-linked polymers may be induced by a number of step-reaction processes or by chain-reaction processes. For the special case of photochemically induced cross-linking it should be noted that whilst chain reactions would be expected to show a very high quantum efficiency for the number of growth steps induced, step-reaction processes are limited to a maximum of one chemical transformation (cross-link, growth step) for each photon absorbed.

During any polymerisation process there exists the possibility that side reactions will create sequences of monomer units pendent to the main chain and known as branches. Branching can occur in many different ways and may involve very short or very long branches and may be frequent or infrequent in relation to the main macromolecular backbone. The formation of cross-links is essentially a special (i.e. extended) case of branching which may, similarly, involve short or long cross-links with a high or low frequency of occurrence along the main polymer backbone. However, whilst the nature and length of the cross-links is not necessarily determined by whether the process is a step reaction or a chain reaction, it is usually much easier to produce long cross-links by chain-reaction polymerisation of olefins and epoxides. In the simplest case if the degree of cross-linking reaches the extent of approximately one cross-link per weight average macromolecule, the result will be an infinite, three-dimensional network macromolecule. The point at which branching and cross-linking form a network is known as the gel point for the system. So far it has been assumed that all units comprising the ultimate macromolecule (linear, branched, or cross-linked) are made up of identical building blocks derived from the same monomer units. If the branching chain is derived from a different monomer or building block, the product macromolecule is more correctly referred to as a graft copolymer. Polymer chain conformations of graft copolymers may be totally different from those of corresponding homopolymers.

Most macromolecules are thermoplastic except for some which decompose below their melting temperatures. This means that there is a temperature range in which they show a transition from a hard glassy material into a soft plastic one. Above this transition the polymer shows varying degrees of viscous flow and rubber elasticity depending on molecular weight and chain structure. Thermoplasticity (flow) is lost through cross-linking. At low degrees of cross-linking, i.e. large separations between the cross-links, the cross-linked material remains rubber elastic above the softening point without being able to flow, whereas increased degrees of cross-linking make the material increasingly hard and brittle. An example of a commercially important lightly cross-linked polymer is vulcanised rubber and very tightly cross-linked polymers are represented by phenol–formaldehyde, melamine–formaldehyde, and phthalic anhydride–glycerol resins. Such resins are infusible and are therefore called thermosetting resins, in contrast to thermoplastic materials.

Another important difference in physical properties between linear (or branched) and cross-linked polymers lies in their solubility. When non-cross-linked polymers are treated with solvents the random coils gradually adopt conformations reflecting the balance between polymer–polymer and polymer–solvent interactions and gradually pass through solvent swollen states into true solutions. Cross-linked polymers are insoluble in all solvents but, according to the degree of cross-linking, they can be swollen to a smaller or larger extent. The degree of swelling may be taken as a direct measure of the degree of cross-linking and it is the insolubility of cross-linked polymers which forms the main basis for the development and applications of polymeric photoresists.[2]

Other ways in which cross-linking affects the properties of any given polymeric system include adhesion and cohesion (to surfaces and to applied liquid inks), permeability to gases and liquids, and refractive index. Several of these are also utilised in imaging and relief plate manufacture.[1,3]

3. FORMATION AND DECAY OF ELECTRONICALLY EXCITED STATES

Before considering details of the ways in which polymers may be photochemically cross-linked it is necessary to have a general framework for understanding light-induced molecular transformations.

Molecules have minimum electronic energy when in the ground state which, for the vast majority of molecules is a singlet state, symbolised S_0. Promotion of an electron from the highest occupied molecular orbital in the ground state to the lowest vacant molecular orbital is the transition of lowest energy which results in a change in electronic configuration. The state produced by this electronic transition may be either singlet or triplet, and will be the first excited state, symbolised S_1 or T_1, respectively. Most light absorption processes obey the Franck–Condon principle and, in consequence, the state obtained is not only electronically but also vibrationally excited. In the gaseous state dissipation of excess vibrational energy will depend on partial and total gas pressures and reactions of 'hot' ground and excited states are often possible. In the solid and liquid states, in contrast, excess vibrational energy is dissipated very rapidly by internal conversion processes leading to the lowest vibrational level of the appropriate electronically excited state.

Singlet states are invariably of higher energy than the corresponding triplet state because of greater electronic repulsion in the former. Excited states of greater energy than S_1 and T_1 (e.g. $S_2, S_3 \ldots T_2, T_3 \ldots$) may be formed by similar excitation processes, but internal conversion of such states to the corresponding first excited states is normally much more rapid in solution than any normal photochemical process and, in consequence, their existence will be largely ignored for purposes of this survey. Since the majority of molecules exist in ground singlet states, it follows, from the rules governing spin conservation, that direct excitation of such molecules must produce mainly singlet excited states. Once formed, the (lowest) excited singlet state may lose its excitation energy in one of four main processes:

(i)   Radiationless conversion back to ground singlet state.
(ii) Radiative conversion back to ground singlet state (fluorescence).
(iii) Quenching of the excited singlet state by interaction with other constituents of the system.
(iv) Conversion to the corresponding triplet excited state (intersystem crossing).

Generation of excited triplet states is normally achieved as a result of (iv) above and, once formed, they may decay by processes analogous to (i)–(iv) with the obvious distinction that radiative decay of triplet states is termed phosphorescence, and that radiationless transition of excited triplet states, back to ground singlet states, involves intersystem crossing.

The various deactivation processes (i)–(iv) listed above are referred to as photophysical processes and a simplified illustration of their inter-dependence is given in Fig. 1.

Quenching of excited states by interaction with other molecules may occur via several mechanisms of which electronic energy transfer, in which large amounts of energy are transferred, is perhaps the most important. Electronic energy transfer may occur for both small molecules and polymer molecules by inter-molecular and intra-molecular processes, the latter being of special significance for polymers since it implies migration of energy along the polymer chains. Such processes are not only of theoretical interest but they have important practical implications, notably in providing methods of stabilising polymers against photodegradation and of achieving controlled degradation by incorporation of 'energy sinks'.

Photophysical processes in small molecules have received much study[4] and four types of electronic energy transfer are recognised:

(a) The 'trivial' type, in which radiation emitted by the excited donor is absorbed by the acceptor. This requires overlap between the relevant emission and absorption spectra and whilst it may operate over large distances, it is not thought to be of general importance in condensed systems.
(b) A non-radiative process involving electron exchange interactions between donor and acceptor, operating over short distances (10–15 Å), with a rate limited by the collision frequency of the interacting species. This is unquestionably the most important energy transfer mechanism and it is governed by quantum

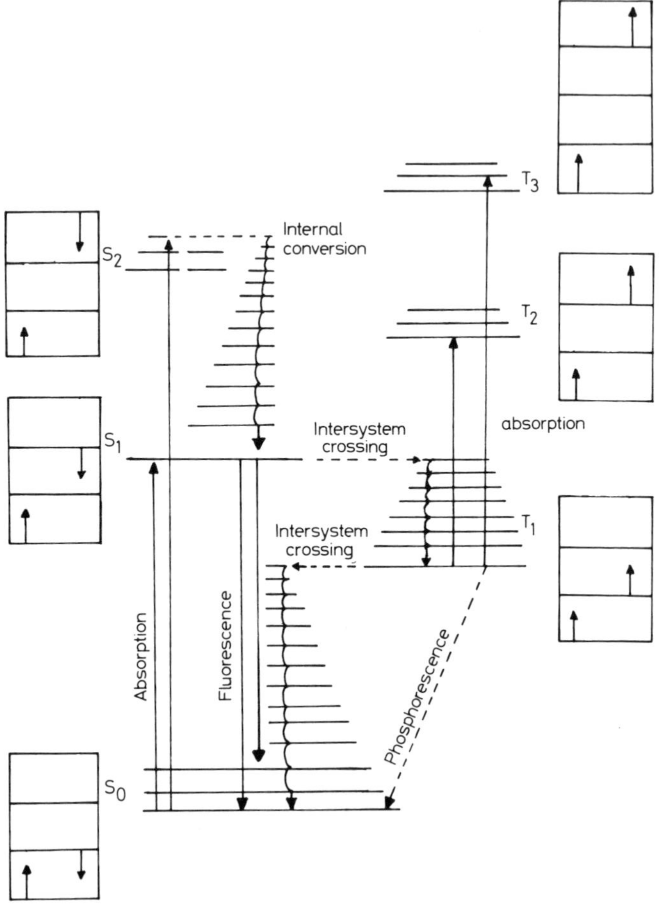

FIG. 1. Energy relationships between electronic and vibrational levels for simple organic molecules.

restrictions which, providing that total spin of the system remains constant, permit a change of spin multiplicity in the interacting species. Typically a triplet excited donor interacts with a singlet ground state acceptor by 'triplet–triplet energy transfer', e.g.

$$\text{Donor }(T_1) + \text{Acceptor }(S_0) \rightarrow \text{Donor }(S_0) + \text{Acceptor }(T_1)$$

(c) A non-radiative process operating over relatively large distances (50–100 Å) involving donor–acceptor dipole–dipole interaction and usually known as the Forster energy transfer mechanism.

Transfer of this type is normally forbidden if there is a change of spin in either partner.

(d) Energy transfer by an exciton mechanism. An 'exciton' is an excited state delocalised over a group of molecules and may be singlet or triplet in character. The molecules must have a high degree of local order and exciton phenomena are therefore most frequently encountered in crystals where the exciton diffusion length may be of the order of 500 Å.

A detailed discussion of photophysical processes for organic molecules[4] is outside the scope of this survey but there are a number of molecular interaction processes, additional to those listed above, which are especially important in photo-excited polymer molecules. Singlet excited states may decay by a process involving interactions with their ground state precursors with intermediate formation of an 'excimer' which, in turn, may emit radiation, undergo intersystem crossing to the triplet level, or condense to give dimeric products.

Triplet excited states frequently undergo depletion via a 'triplet–triplet annihilation mechanism' which results in the formation of an excited singlet molecule and a ground state singlet molecule, e.g.

$$M(T_1) + M(T_1) \rightarrow M(S_1) + M(S_0)$$

For polymers it is important to remember that excimer formation and triplet–triplet annihilation may be intra-molecular or inter-molecular depending on the temperature and physical state of the system.[5]

4. FORMATION OF IONIC AND DIPOLAR INTERMEDIATES BY LIGHT-INDUCED PROCESSES

Direct ionisation of most organic molecules is a relatively high-energy process not normally encountered in photo-induced reactions. There are however two types of photochemical reactions which may, under appropriate circumstances, lead to either dipolar intermediates or fully charged molecular species. These are, respectively, well defined photo-chemical reactions of charge-transfer complexes, and various photo-induced redox reactions. However, the recently characterised role of 'exciplexes' has helped to minimise apparent differences between the two types of photochemical reactions, as will become apparent in the subsequent discussions.

According to Mulliken,[6] charge-transfer complexes arise from interactions between donor molecules having high-energy filled orbitals (i.e. low ionisation potentials, I_D), and acceptors having low-energy unfilled orbitals (i.e. high electron affinities, E_A), viz.:

$$D + A \rightleftharpoons [D, A \leftrightarrow D^{\cdot +}, A^{\cdot -}] \xrightarrow{h\nu} [D^{\cdot +}, A^{\cdot -} \leftrightarrow DA]$$

ground state   excited state

In contrast to Lewis acid–Lewis base adducts, the ground state of a charge-transfer complex may be considered as a resonance hybrid with only a small contribution from the canonical structure representing electron transfers between D and A (i.e. $D^{\cdot +}A^{\cdot -}$). A low-lying electronically excited state of the complex would have a major contribution from the electron transfer canonical form ($D^{\cdot +}A^{\cdot -}$) and explains the appearance of new absorption bands on mixing D and A. An important justification for this theory of charge-transfer spectra is the general applicability of empirical relationships between ionisation potential of the donor (I_D), electron affinity of the acceptor (E_A), and the energy of the charge-transfer transition, which may be generalised as:

$$h\nu = I_D - E_A + C$$

where C is a constant representing Coulombic forces. A full description of this topic is given in the review by Davidson.[7]

A useful discussion of photochemical effects involving charge-transfer transitions requires a somewhat more detailed description of possible excited states which contribute to the bonding forces. In particular, higher excited charge-transfer states and locally excited states of the 'no bond' structure (e.g. D*A and DA*) should be considered so as to permit more easily an integration of photochemical and photophysical properties of molecular complexes stable in their ground states (charge-transfer complexes), and those stable only in their excited states. The latter are most frequently referred to as exciplexes and were first demonstrated to have practical value by Weller and his associates.[8] For present purposes it is important to recognise that excited states of charge-transfer complexes may arise either by direct excitation of the ground state complex, or by diffusional controlled collision between locally excited donor (or acceptor) and ground state acceptor (or donor), viz. Fig. 2. Exciplexes are represented as $D^* \cdots A$ and $D \cdots A^*$ and the subscripts (s) are used to indicate a degree of solvation equilibrium around the particular species (F–C and REL imply Franck–Condon and solvent relaxed excited states respectively); a further complication arises because of the possibility for intersystem crossing to produce corresponding triplet excited states for

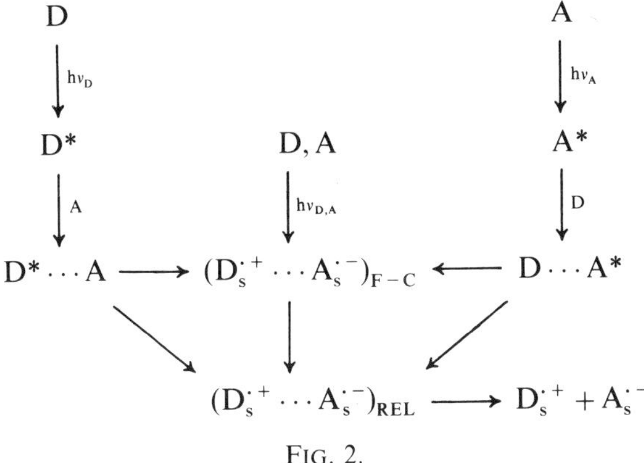

FIG. 2.

any, or all, of the intermediates. For a full discussion of these very important photophysical transitions, the reader is referred to an enlightening survey by Ottolenghi,[9] which is all the more valuable for a clear exposition of the way in which locally excited states of D, A, D^+ and A^- may arise from higher excitation of the lowest charge-transfer state. However, the most important aspect of the scheme illustrated above is the clear manner in which it provides an understanding of the possible role of solvent polarity in determining the outcome of reactions involving exciplexes or charge transfer complexes.[7]

5. PHOTO-CROSS-LINKING BY STEP-REACTION PROCESSES

Photo-cross-linking of polymers is one of the competing processes involved in photodegradation of polymers in both the solid state and in solution.[10] Frequently precise details of the chemical transformations are not fully understood but increased mechanical strength and weather resistance are observed for low degrees of cross-linking while excessive brittleness and loss of mechanical strength may accompany higher degrees of cross-linking. These processes are best exemplified by the case of polyethylene where it is quite feasible to improve some mechanical and thermal properties by photo-cross-linking in the presence of sensitisers such as benzophenone (P_n—H represents any carbon–hydrogen bond in polyethylene):

$$(C_6H_5)_2C{=}O \xrightarrow{h\nu} (C_6H_5)_2C{=}O(S_1) \longrightarrow (C_6H_5)_2C{=}O(T_1)$$

$$(C_6H_5)_2C{=}O(T_1) + P_n\text{—}H \longrightarrow (C_6H_5)_2\dot{C}\text{—}OH + \dot{P_n}$$

$$\dot{P_n} + \dot{P_n} \longrightarrow P_n\text{—}P_n \text{ (cross-linked polymer)}$$

Greatly improved efficiency in cross-linking of plastics such as polyethylene may be accomplished by use of polymerisable aromatic ketone sensitisers such as *p*-acryloyloxybenzophenone (**I**). In this case radical sites formed on

$$CH_2{=}CHCOO\!\!-\!\!\bigcirc\!\!-\!\!\overset{\overset{\textstyle O}{\|}}{C}\!\!-\!\!\bigcirc \qquad\qquad \mathbf{I}$$

the polymer backbone by hydrogen abstraction may initiate polymerisation of the acrylic monomer, with ultimate formation of longer cross-links between the polyethylene chains following the expected radical recombination processes. This, of course, is a rather special case of cross-linking by chain reactions to be discussed later (section 6). In principle these types of photo-cross-linking can be applied to any preformed polymer having C—H bonds attached to either the backbone or to side groups and branches. More usually, however, where photo-cross-linking is an essential part of a technologically important process, such as a photoresist, it is preferable to employ one of the two following methods of stepwise cross-linking.

Organic azides (RN_3) afford a highly convenient source of monovalent nitrogen species which are highly reactive and known as nitrenes ($R\ddot{N}{:}$). Formation of nitrenes may be accomplished by both thermal and photochemical activation of azido compounds and, for practical reasons, the most important precursors are aromatic azides (ArN_3), arylsulphonyl azides ($ArSO_2N_3$), and azidoformates ($ROCON_3$). In all cases the azido compound yields (initially) a singlet nitrene which may then undergo intersystem crossing to the corresponding triplet state.[11] For the case of light-induced decomposition of azides, it is possible to generate triplet nitrenes directly by the usual expedient of using triplet sensitisers for the photolysis.

$$R{-}N_3 \xrightarrow[\text{or } \Delta]{h\nu} R\ddot{N}{:} + N_2 \longrightarrow R\ddot{\ddot{N}}{\cdot}$$

$$\qquad\qquad\qquad \text{singlet} \qquad\qquad\qquad \text{triplet}$$

Different reaction paths are available for deactivation of singlet and triplet nitrenes:

(a) Recombination of nitrenes is spin-allowed for both singlet and triplet states.

$$2R{-}\ddot{N}{:} \longrightarrow R{-}N{=}N{-}R$$

(b) Electrophilic attack on bonding electron pairs can occur by insertion into a C—H bond when the nitrene is in a singlet state, or by hydrogen abstraction (possibly the most general reaction of nitrenes) when the nitrene is in the triplet state.

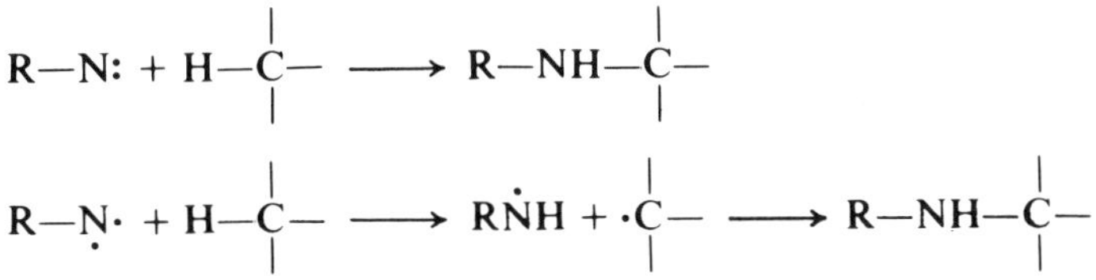

$$R-N{:} + H-\overset{|}{\underset{|}{C}}- \longrightarrow R-NH-\overset{|}{\underset{|}{C}}-$$

$$R-\overset{\cdot}{\underset{\cdot}{N}}H + H-\overset{|}{\underset{|}{C}}- \longrightarrow R\dot{N}H + \cdot\overset{|}{\underset{|}{C}}- \longrightarrow R-NH-\overset{|}{\underset{|}{C}}-$$

(c) Addition to double bonds is an important reaction for both singlet and triplet nitrenes.

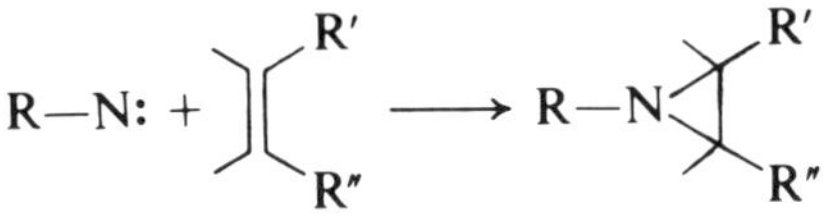

$$R-N{:} + \quad \longrightarrow R-N$$

The very high and relatively unselective reactivity of nitrenes, coupled with the convenience of synthesis of appropriate azido compounds, affords a very large number of ways in which polymers may be conveniently cross-linked.[2,12] Most important are the use of bis-azido compounds (N_3—R—N_3) which will induce cross-linking in preformed polymers containing saturated and/or unsaturated hydrocarbon groupings by the various C—H insertion and C=C addition possibilities, and the use of polymers functionalised with pendant azido groups which may then undergo a variety of photo-induced self-condensation processes. Representative examples of difunctional azido compounds and polymers having azido substituents are given (structures **II**–**VI**). The presence of conjugated carbonyl groups in

$$N_3COO(CH_2)_4OCON_3$$

II

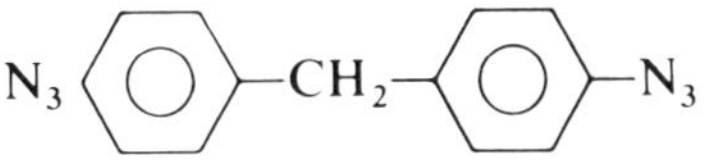

III

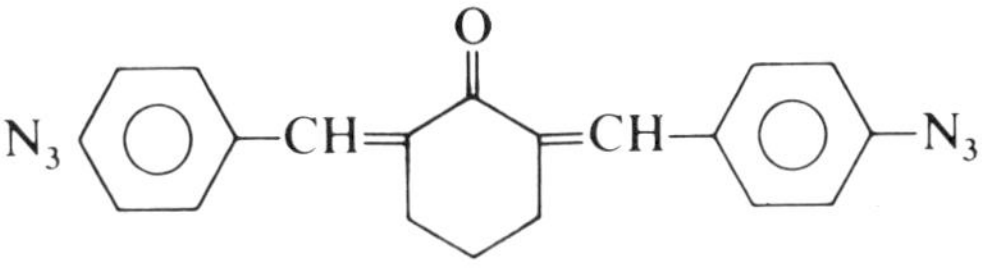

IV

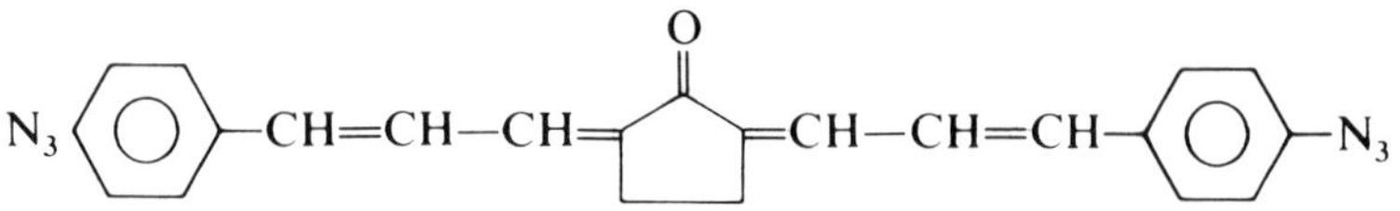

V

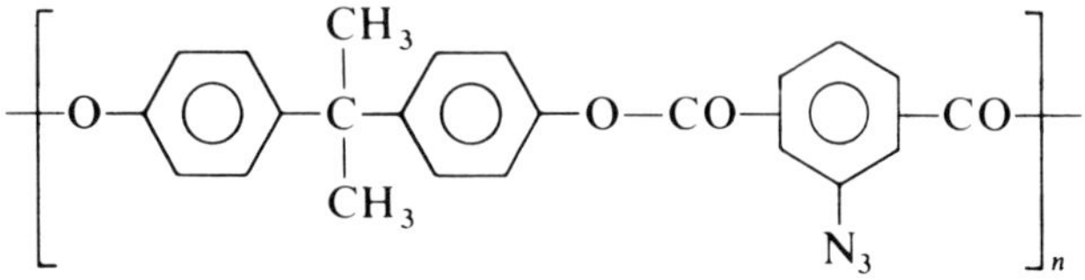

VI

any of these structures serves both to increase the wavelength sensitivity of the azido derivative and to function as an intra-molecular sensitising unit. It is also feasible to introduce pendant azido groups by copolymerisation of suitable functional monomers such as vinyl(p-azidobenzoate).

A second, even more widely employed method of photo-cross-linking via a step reaction utilises the well known propensity of cinnamic acid and its derivatives to undergo photo-induced cyclodimerisation.[2,12]

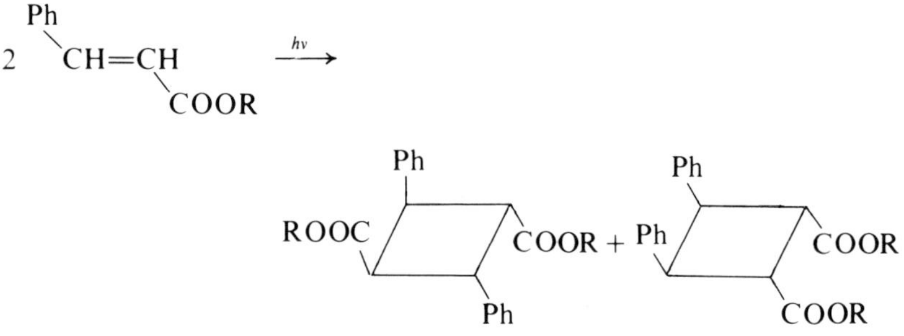

(+ other isomers)

The reaction is efficient, in a photochemical sense, especially when sensitised with a very wide range of aromatic carbonyl compounds and organic dye molecules. Triplet–triplet energy transfer is presumed to be the most important type of sensitisation but in some cases, because of the relative triplet energy levels, other mechanisms (e.g. exciplex intermediates) must prevail. As with many other α,β-unsaturated carbonyl compounds, cyclodimerisation of cinnamic acid esters may also be accomplished by direct excitation of the lowest lying singlet states. Cinnamic acid moieties are readily introduced into polymers by reactions of side chain OH groups, e.g. in poly(vinyl alcohol), with cinnamoyl chloride and the generalised cross-linking mechanism may be represented as follows:

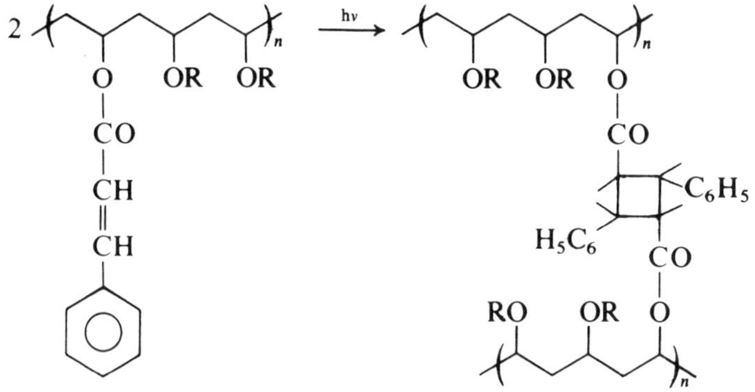

$$R = H \text{ or } -CO-CH=CH-C_6H_5$$

Increased reactivity in cross-linking is observed for vinyl homologues of cinnamic acid such as may be obtained from appropriate reactions of cinnamylideneacetyl chloride, $Ph(CH=CH)_2COCl$. Free-radical polymerisation of vinyl cinnamate and related molecules proceeds with substantial degrees of cyclopolymerisation and consequent loss of cyclodimerisable substituent groups. On the other hand, cationic polymerisation of appropriately functionalised cinnamate and the related chalcone monomers proceeds without consumption of the cyclodimerisable units. Styryl and alkyl vinyl ether functionalised molecules have been most usefully employed for this purpose, as exemplified by compounds **VII, VIII** and **IX**.

Cyclodimerisation of other types of unsaturated molecule may also be utilised to effect cross-linking of preformed polymers and, here also, all

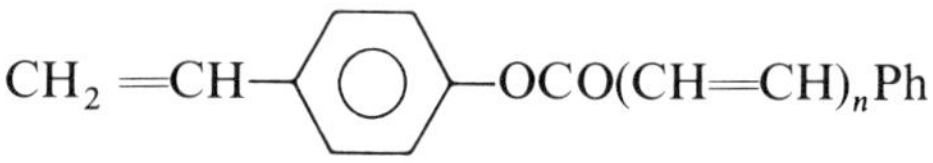

$$CH_2 =CH- \bigcirc -OCO(CH=CH)_nPh \qquad n = 1, 2$$

VII

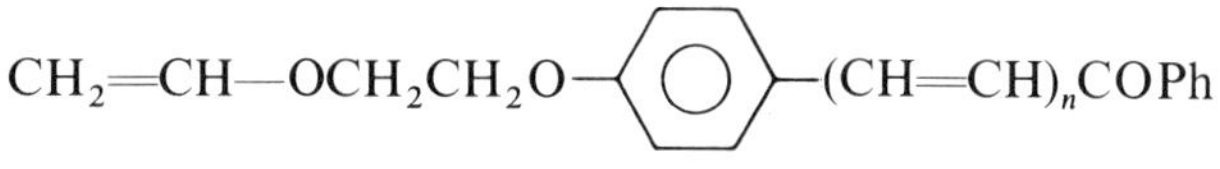

$$CH_2=CH-OCH_2CH_2O- \bigcirc -(CH=CH)_nCOPh$$

VIII

$$CH_2=CHOCH_2CH_2OCO(CH=CH)_n-Ph$$

IX

A. LEDWITH

FIG. 3. Methods of introducing photo-active units for cross-linking by step reactions.

methods of polymer synthesis are applicable to formation of photo-chemically active polymers. In addition to the cinnamates already mentioned, other groups useful in photodimerisation include chalcones (RCH=CHCOAr), strained cyclo-olefins

$$\left(\text{ROCOCH}\begin{array}{c}\diagup\text{CPh}\\ \parallel\\ \diagdown\text{CPh}\end{array}\right)$$

stilbenes (ArCH=CHAr′), coumarins (a special class of α,β-unsaturated ester), anthracenes and maleimides.

Functional groups such as those described above and active in photo-induced step reactions may be utilised for polymer cross-linking in three distinct ways: small molecule addenda, main-chain incorporated, and appended substituents, as indicated in Fig. 3.

6. PHOTO-CROSS-LINKING BY CHAIN-REACTION PROCESSES

Cross-linking by chain processes is very easily accomplished in polymer systems and most commonly involves free-radical intermediates.[13,14] The necessary vinyl functionality may arise in one of the following ways:

(i)   Substituted ethylenic groups in the polymer backbone such as would be obtained in polyesters derived from step reactions of maleic and fumaric acid derivatives and chain reactions of 1,3-conjugated dienes, e.g.

$$\sim\!\!\text{ROCOCH}\!=\!\text{CHCOOR}\!\sim\quad cis\text{- and }trans\text{-forms}$$

$$\sim\!\!\text{CH}_2\!-\!\overset{\overset{\textstyle R}{|}}{\text{C}}\!=\!\text{CH}\!-\!\text{CH}_2\!\sim\quad cis\text{- and }trans\text{-forms}$$

$$R = H, CH_3, Cl$$

(ii)   Pendant unsaturated units derived *either* by selective polymerisation of monomers having more than one type of vinyl group *or* by post-polymerisation functionalisation of appropriate polymeric structures, e.g.

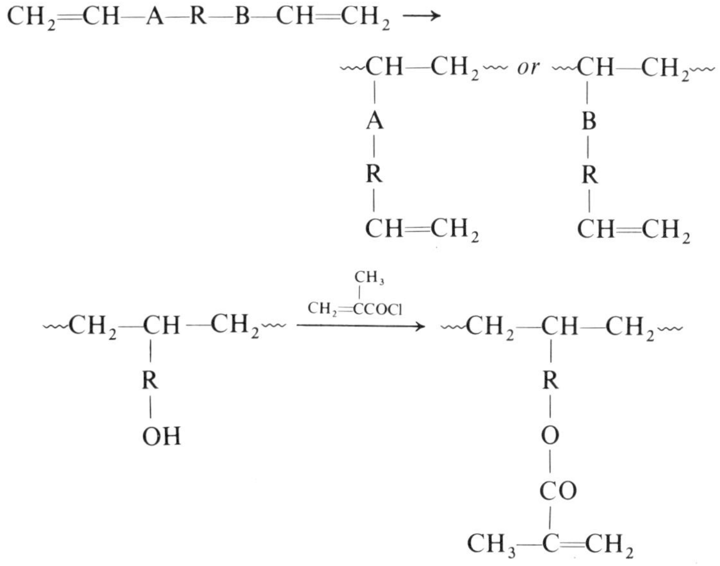

$$CH_2{=}CH{-}A{-}R{-}B{-}CH{=}CH_2 \longrightarrow$$

$$\begin{array}{ccc}
\text{\textasciitilde}CH{-}CH_2\text{\textasciitilde} & or & \text{\textasciitilde}CH{-}CH_2\text{\textasciitilde} \\
| & & | \\
A & & B \\
| & & | \\
R & & R \\
| & & | \\
CH{=}CH_2 & & CH{=}CH_2
\end{array}$$

$$\text{\textasciitilde}CH_2{-}\underset{\underset{OH}{|}}{\underset{|}{CH}}{-}CH_2\text{\textasciitilde} \xrightarrow{\overset{\overset{CH_3}{|}}{CH_2{=}CCOCl}} \text{\textasciitilde}CH_2{-}CH{-}CH_2\text{\textasciitilde}$$

with the product bearing

$$\begin{array}{c}
R \\
| \\
O \\
| \\
CO \\
| \\
CH_3{-}C{=}CH_2
\end{array}$$

(iii)   Polymerisation of monomeric species which have more than one polymerisable group, e.g.

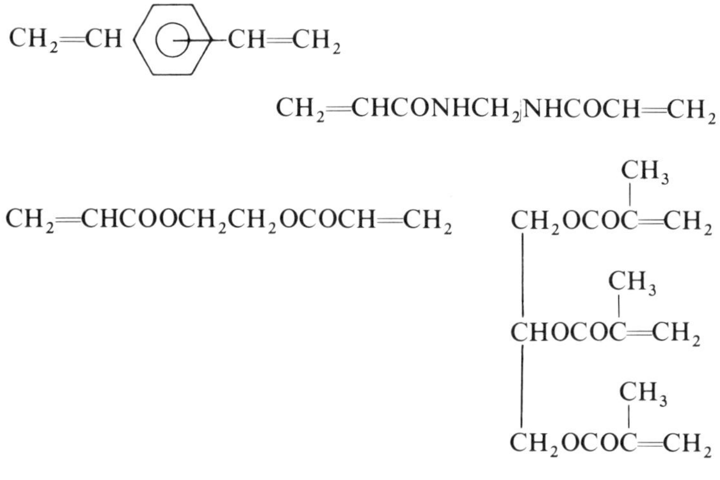

$$CH_2{=}CH{-}\langle\bigcirc\rangle{-}CH{=}CH_2$$

$$CH_2{=}CHCONHCH_2NHCOCH{=}CH_2$$

$$CH_2{=}CHCOOCH_2CH_2OCOCH{=}CH_2 \qquad CH_2OCO\overset{\overset{CH_3}{|}}{C}{=}CH_2$$

$$\begin{array}{c}
| \\
CHOCO\overset{\overset{CH_3}{|}}{C}{=}CH_2 \\
| \\
CH_2OCO\overset{\overset{CH_3}{|}}{C}{=}CH_2
\end{array}$$

                         A. LEDWITH

In these cases the polymerisation will yield cross-linked (i.e. network) polymers directly and most practical applications are based on this principle. A typical network is illustrated in Fig. 4. This diagram is schematic only and in practice network formation will be quantitatively affected by cyclisation and diffusion phenomena as described in Chapter 4.

Any of the reagents normally used to accomplish photochemical initiation of free-radical polymerisation may, in principle, be applied to

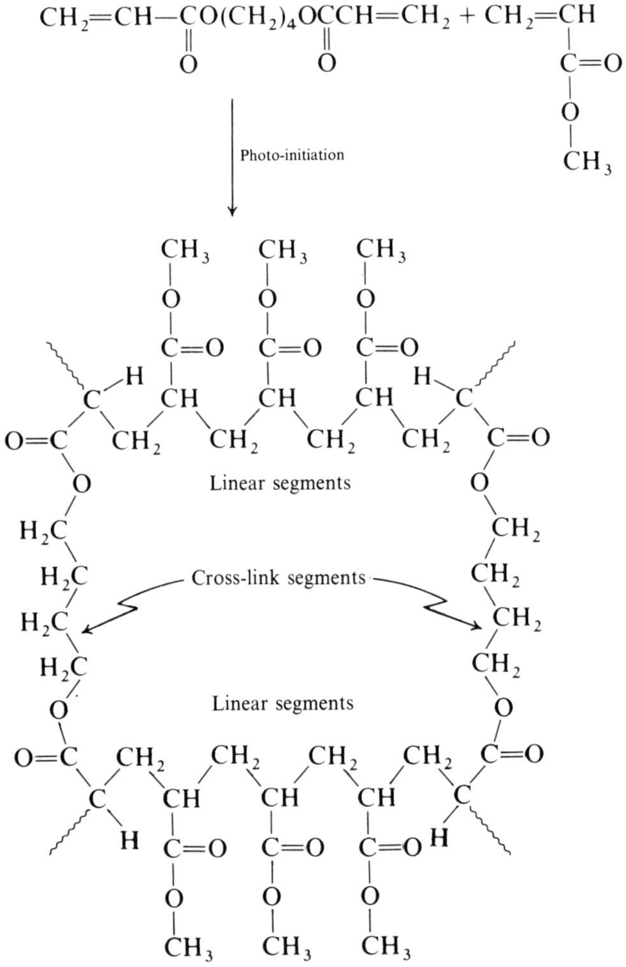

$$CH_2{=}CH{-}\underset{\underset{O}{\|}}{C}O(CH_2)_4O\underset{\underset{O}{\|}}{C}CH{=}CH_2 + CH_2{=}\underset{\underset{\underset{\underset{CH_3}{|}}{O}}{\underset{|}{C}{=}O}}{CH}$$

$$\downarrow \text{ Photo-initiation}$$

FIG. 4. Cross-link formation in a typical photochemically active polymerising system.

initiation of cross-linking for all possibilities listed above but practical considerations restrict the choice to two types of photo-initiation.[15]

(a)   Photofragmentation of benzoin or acetophenone derivatives, e.g.

$$\text{PhCO—}\overset{\displaystyle \overset{\textstyle OMe}{|}}{\underset{\displaystyle \underset{\textstyle OMe}{|}}{C}}\text{—Ph} \xrightarrow{h\nu} \text{Ph}\dot{C}O + \cdot\overset{\displaystyle \overset{\textstyle OMe}{|}}{\underset{\displaystyle \underset{\textstyle OMe}{|}}{C}}\text{—Ph}$$

(2,2-dimethoxy-
2-phenylacetophenone)

$$\text{PhCOCH(OEt)}_2 \xrightarrow{h\nu} \text{Ph}\dot{C}O + \cdot\text{CH(OEt)}_2$$

(2,2-diethoxy-
acetophenone)

$$\text{PhCOCH—PH} \xrightarrow{h\nu} \text{Ph}\dot{C}O + \cdot\text{CHPh}$$

with O_iPr substituents

(benzoin isopropyl
ether)

(b)   Formation of free radicals via exciplex interactions of aromatic carbonyl compounds with a wide range of amino compounds—especially tertiary amines, e.g.[16]

$$\text{Ar}_2\text{C}=\text{O} \xrightarrow{h\nu} \longrightarrow \text{Ar}_2\text{C}=\text{O}(T_1) \xrightarrow{RCH_2NR_2}$$

$$(\text{Ar}_2\text{C}=\text{O}, \text{RCH}_2\text{NR}_2)^*(T_1)$$

exciplex

$$(\text{Ar}_2\text{C}=\text{O}, \text{RCH}_2\text{NR}_2)^* \longrightarrow$$

$$\text{Ar}_2\text{C}=\text{O}^{\overset{-}{\cdot}}\text{RCH}_2\overset{+\cdot}{\text{N}}\text{R}_2 \longrightarrow \text{Ar}_2\dot{\text{C}}\text{—OH} + \text{R}\dot{\text{C}}\text{HNR}_2$$

In this case it is the amine-derived radical which initiates; the semipinacol radicals derived from the carbonyl compounds are known to form pinacol dimers very readily.

Initiating systems in class (a) are most useful in clear polymerising systems such as transparent surface coatings. In contrast, ketone/amine combinations are especially advantageous in pigmented polymerising systems such as are found in inks and many decorative coatings. The reasons for this lie mainly in the increased wavelength sensitivity (to longer wavelength) of the ketone employed, permitting light activation in the presence of TiO_2 pigments. Titanium dioxide is perhaps the most common ingredient of all pigmented coatings and absorbs u.v. light up to approximately 390 nm. Ketones active in photo-initiation (benzil, thioxanthone, fluorenone) have long-wavelength absorption tails extending beyond 400 nm.

A very special component of ketone/amine initiation systems is 4,4'-bis(N,N-dimethylamino)benzophenone (Michler's ketone) which can serve as both carbonyl compound and amine activator, and hence may be used by itself or (more efficiently) in conjunction with benzophenone.

In all cases of photo-induced (chain) cross-linking it is preferable to utilise mixtures of preformed polymers, with in-chain or pendant polymerisable groups, with monofunctional and polyfunctional olefins. Alternatively network structures may be derived by homopolymerisation of polyfunctional monomers or mixtures of monofunctional and polyfunctional monomers. Monomer reactivity ratios will control the extent of incorporation of the various polymerisable groups present in gel systems and it is clear that the extent of cross-linking will be largely determined by the degree of reactivity (and hence degree of incorporation) of any polyfunctional monomer present in the system. A further influence on the degree of cross-linking arises from the known differences in mechanisms of termination for the usual free-radical polymerisations. This follows since free radicals may terminate in a bimolecular manner by either combination or disproportionation, e.g.

$$
\begin{array}{c}
\overset{\displaystyle R}{\underset{|}{}} \quad \overset{\displaystyle R}{\underset{|}{}} \\
\sim\!\!\text{CH}_2\text{CH}\!-\!\text{CH}\!-\!\text{CH}_2\!\!\sim
\end{array}
$$

$$
2\sim\!\!\text{CH}_2\!-\!\overset{\displaystyle \cdot}{\text{C}}\text{HR} \quad
\begin{array}{c}
\overset{k_c}{\nearrow} \quad \text{combination}\\[2mm]
\underset{k_d}{\searrow}
\end{array}
$$

$$
\sim\!\!\text{CH}\!=\!\text{CHR} + \sim\!\!\text{CH}_2\text{CH}_2\text{R}
$$

disproportionation

Clearly combination will favour formation of cross-links and the ratio k_c/k_d for any given class of polymerising unit, at a particular temperature, will play an important role in determining the nature and extent of polymer cross-links. Typically, for room temperature polymerisations, styrene-derived radicals ($\sim\!\!\text{CH}_2\!-\!\overset{\displaystyle \cdot}{\text{C}}\text{HPh}$) terminate exclusively by combination and hence styrene is commonly employed in cross-linking systems. On the other hand, radicals derived from methacrylic acid

$$\left(\text{---CH}_2\text{---}\overset{\overset{\displaystyle CH_3}{\displaystyle |}}{\underset{\underset{\displaystyle COOR}{\displaystyle |}}{C}}\cdot\right)$$

terminate by combination to approximately 30 % at room temperature, and exhibit exclusively disproportionation at temperatures above about 70 °C.

6.1. Cross-linking by Thiol–Olefin Reactions

Alkane thiols and aliphatic olefins readily undergo photochemically induced addition reactions.[17] Benzophenone is one of the best sensitisers for the process and the overall reactions may be represented as follows:

$$(Ph_2C{=}O)T_1 + RSH \longrightarrow Ph_2\dot{C}{-}OH + RS\cdot$$

$$2\,Ph\dot{C}{-}OH \longrightarrow Ph_2\overset{\overset{\displaystyle OH}{\displaystyle |}}{C}{-}\overset{\overset{\displaystyle OH}{\displaystyle |}}{C}Ph_2$$

$$RS\cdot + CH_2{=}CHR_1 \longrightarrow RS{-}CH_2{-}\dot{C}HR_1$$

$$RSCH_2\dot{C}HR_1 + RSH \longrightarrow RSCH_2CH_2R_1 + RS\cdot, \text{ etc.}$$

Overall the process is a chain reaction in which the elements of RSH are added across the vinyl group of $R_1CH{=}CH_2$. These reactions lend themselves well to the design of cross-linkable polymer systems since polyfunctional thiols, polyfunctional olefins, thiol functionalised polymers and olefin-containing polymers are readily available. It should be noted, however, that although the overall sequence is a chain reaction, each growth process involves a single (i.e. step) addition of, for example, RSH to a double bond.

6.2. Photochemically Induced Ionic Chain Processes

In contrast to free-radical systems, only a small number of olefins and strained hetero-cyclic molecules are susceptible to ionic polymerisations.[18]

Anionic reactions have not so far found any application in photo-initiated cross-linking but cationic polymerisations have become increasingly important in recent years. Cationic polymerisability is found in olefins having electron-releasing substituents, e.g. $ROCH=CH_2$, $PhCH=CH_2$, and in strained cyclic ethers such as epoxides. Of these, only epoxides have found significant application in cross-linkable systems—especially as utilised in surface coatings—but this may reflect the availability of multifunctional epoxides from the widely used epoxy resin industry.[19]

Polymerisation of an epoxide by cationic processes may be represented as follows:

$$X^{+} + RCH\!-\!CH_2 \longrightarrow RCH\!-\!CH_2 \xrightarrow{\ RCH-CH_2\ }$$

$$X\!-\!OCH_2CH\!-\!\overset{+}{O}\underset{CHR}{\overset{CH_2}{\diagup\diagdown}} \xrightarrow{\ RCH-CH_2\ } XOCH_2\overset{R}{\overset{|}{C}}HOCH_2\overset{R}{\overset{|}{C}}H\!-\!\overset{+}{O}\underset{CHR}{\overset{CH_2}{\diagup\diagdown}},\ \text{etc.}$$

In the absence of deliberately added nucleophilic agents, there need not be any real termination and the propagating oxonium ion is relatively uninfluenced by atmospheric oxygen. This is one of the reasons why photo-initiated epoxide cross-linking is considered to be potentially better than free-radical systems where atmospheric oxygen may exhibit a marked retarding effect. Epoxide polymerisations are, however, subject to a variety of chain-transfer reactions of which transfer to a protic molecule is the most likely, e.g.

$$\sim\!\!\overset{+}{O}\underset{CHR}{\overset{CH_2}{\diagup\diagdown}} + R_1OH \longrightarrow \sim\!\!OCH_2\overset{R}{\overset{|}{C}}H\!-\!OR_1 + H^{+}$$

$$H^{+} + RCH\!-\!CH_2 \longrightarrow H\!-\!\overset{+}{O}\underset{CHR}{\overset{CH_2}{\diagup\diagdown}},\ \text{etc.}$$

Transfer processes are not particularly disadvantageous when cross-linking is required and hence the use of difunctional epoxides and epoxide-substituted polymers affords the opportunity for efficient cross-linking, e.g.

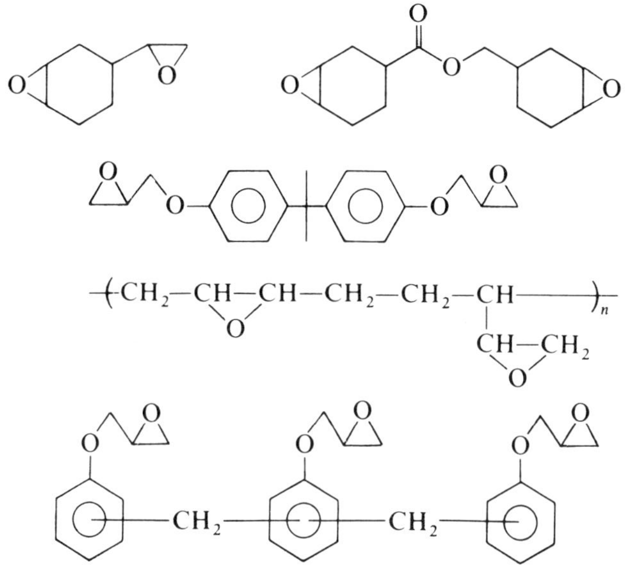

It is also highly convenient to synthesise polymers having pendant epoxide groups by free-radical copolymerisation of glycidyl methacrylate,

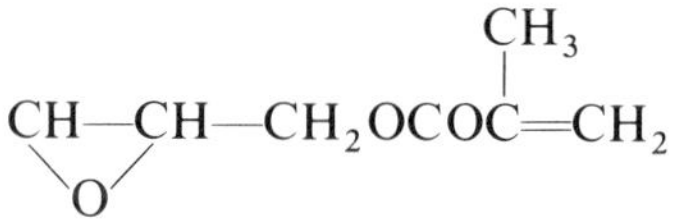

$$CH\!-\!CH\!-\!CH_2OCOC\!\!=\!\!CH_2$$
$$(CH_3)$$

Initiation requires the formation of either protonic acid, carbocation, or Lewis acid species and this may be accomplished by photochemical decomposition of aryldiazonium (ArN_2^+), diaryliodonium (Ar_2I^+), and triarylsulphonium (Ar_3S^+) salts.[19–21] Each salt must have an appropriate anion (e.g. PF_6^-, AsF_6^-, BF_4^-) and the detailed chemistry of the photofragmentation is complicated and different for the three types of salt. The best-known case (aryldiazonium salts) appears to yield free Lewis acid on photolysis which then initiates the epoxide polymerisation in the well established manner, e.g.

$$ArN_2^+\,BF_4^- \xrightarrow{h\nu} ArF + N_2 + BF_3$$

For the other types of salt, acid fragments derived in part from monomer, solvents, etc., may be the important intermediates but detailed mechanistic information is lacking. A suggested[20,21] mechanism is illustrated with reference to photodecomposition of triarylsulphonium cation ($RH =$ solvent or monomer)

$$Ar_3S^+ \xrightarrow{h\nu} (Ar_3S)^*$$
$$\text{singlet or triplet}$$

$$(Ar_3S^+)^* \longrightarrow Ar_3S^{+\cdot} + Ar^{\cdot}$$

$$Ar_2S^{+\cdot} + RH \longrightarrow Ar_2S + H^+ + R^{\cdot}$$

$$H^+ + \text{monomer (epoxide)} \longrightarrow \text{polymer}$$

Sulphonium and iodonium salts have absorption spectra in the ultraviolet region only ($\lambda_{max} \sim 250\,\text{nm}$) and hence long-wavelength-absorbing sensitisers must be utilised for maximum efficiency with most commercially available light sources. Typical sensitisers include dyes such as Acridine Orange, aromatic molecules such as substituted anthracenes and perylene, and a number of aromatic ketones such as thioxanthone.[22] It is thought that the sensitiser functions by donation of an electron to the onium salt via exciplet formation. For example,

$$\text{Sens} \xrightarrow{h\nu} \text{Sens}^*$$
$$\text{singlet or triplet}$$

$$\text{Sens}^* + Ar_3S^+ \longrightarrow (\text{Sens}, Ar_3S^+)^*$$

$$(\text{Sens}, Ar_3S^+)^* \longrightarrow \text{Sens}^{+\cdot} + Ar_2S + Ar^{\cdot}$$

$$\text{Sens}^{+\cdot} + RH \longrightarrow \text{Sens} + R^{\cdot} + H^+$$

In a rather novel extension of this type of mechanism it has been demonstrated that typical u.v. free-radical initiators such as 2,2-dimethoxy-2-phenylacetophenone may act as sensitisers for onium salt decomposition by electron-transfer reactions of the initially formed free radicals.[23]

Iodonium salts are more active than sulphonium salts in this type of electron-transfer process and initiation of polymerisation is assumed to arise via direct reaction of product carbocation with the epoxide monomer.

7. APPLICATIONS OF PHOTO-CROSS-LINKABLE POLYMER SYSTEMS

The advantages of light-induced cross-linking have led to the use of photo-active polymer systems in a wide variety of applications, and new possibilities are continually suggested.

Perhaps the oldest known application of photochemically induced cross-linking is in the design and use of 'photoresists'.[1] Cross-linking photoresists are synthetic varnishes or coatings which become insoluble on irradiation with actinic light. By exposing a resist coating to an optical pattern, and washing off the soluble (i.e. unexposed) parts of the film after exposure, polymeric images can be produced on a variety of substrates. In outline, the mechanism of resist action is well understood. The unexposed material is either a soluble (linear) polymer or a monomer composition. It is applied to the substrate in the form of a thin film. On irradiation of the film cross-links are produced which connect either preformed polymer chains or the emerging macromolecules, until a three-dimensional network, or gel, is formed. The polymeric image is utilised as a protective coating during etching or other surface treatment of the exposed substrate. Printed circuits are manufactured in this way and integrated circuits on silicon chips are also formed via photoresists although the line widths are very much smaller. An increasingly important application of photoresists is in the manufacture of printing plates, supplanting the more traditional methods involving molten metals.

Another major application of photo-cross-linking is in the curing of a wide variety of surface coatings. Examples include wood finishing, record sleeve covers, vinyl floor coatings, beverage cans, and the coating of optical fibres. The coatings may be pigmented or clear and can have protective and decorative functions.

Ultraviolet-sensitive printing inks are increasing in importance because the use of 'completely reactive' fluids offers both economic and environmental benefits in the curing (i.e. hardening) processes.[14] These inks are made up of pigments dissolved in completely reactive monomer–polymer systems which harden without loss of solvent to the atmosphere, and without heating requirements. Very high cure speeds have been accomplished in commercial systems (e.g. < 0.01 s) and the technique offers the additional advantages of immediate shape cutting and stacking.

For the future it is clear that photochemical activity in polymer-based systems will play an increasingly important role in methods of printing, data storage and retrieval, and surface treatments.

REFERENCES

1. DeForest, W. S., *Photoresist, Materials and Processes*, New York, McGraw-Hill, 1975.
2. Reiser, A., *J. Chim. Phys.*, **77** (1980), p. 469.
3. Brinckman, E., Delzenne, G., Poot, A. and Willems, J., *Unconventional Imaging Processes*, London, Focal Press, 1978; Jacobson, K. I. and Jacobson, R. E., *Imaging Systems*, London, Focal Press, 1976.
4. Torro, N. J., *Modern Molecular Photochemistry*, Menlo Park, California, Benjamin–Cummings, 1978.
5. Holden, D. A. and Guillet, J. E., in *Developments in Polymer Photochemistry*, Volume 1, Allen, N. S. (Ed.), London, Applied Science, 1980, p. 27.
6. Mulliken, R. S., *J. Amer. Chem. Soc.*, **74** (1952), p. 811.
7. Davidson, R. S., in *Molecular Association*, Volume 1, Foster, R. (Ed.), London, New York, Academic Press, 1975, p. 216.
8. Leonhardt, H. and Weller, A., *Ber. Bunsenges. Phys. Chem.*, **67** (1963), p. 791; Weller, A., *Pure Appl. Chem.*, **16** (1968), p. 115.
9. Ottolenghi, N., *Acc. Chem. Res.*, **6** (1973), p. 153.
10. McKellar, J. F. and Allen, N. S., *Photochemistry of Man-Made Polymers*, London, Applied Science, 1979; Ranby, B. and Robek, J. F., *Photodegradation, Photo-oxidation and Photostabilisation of Polymers*, London, Wiley, 1975.
11. Reiser, A. and Wagner, H. M., in *The Chemistry of the Azido Group*, Patai, S. (Ed.), New York, Interscience, 1971, p. 461.
12. Delzenne, G., in *Encyclopaedia of Polymer Science and Technology*, Suppl. Volume 1, Bikales, N. F. (Ed.), New York, Wiley, 1976, p. 401.
13. Ledwith, A., *Pure Appl. Chem.*, **49** (1977), p. 431; *J. Oil Col. Chem. Assoc.*, **59** (1976), p. 157.
14. Pappas, S. P. (Ed.), *U.V. Curing: Science and Technology*, Stamford, Connecticut, Technology Marketing Corp., 1978.
15. McGinniss, V. D., *Phot. Sci. Eng.*, **23** (1979), p. 124.
16. Ledwith, A., Bosley, J. A. and Purbrick, M., *J. Oil Col. Chem. Assoc.*, **61** (1978), p. 95.
17. Morgan, C. R., Magnotla, F. and Ketley, A. D., *J. Polym. Sci., Polym. Chem.*, **15** (1977), p. 627.
18. Jenkins, A. D. and Ledwith, A. (Eds.), *Structure, Reactivity and Mechanism in Polymer Chemistry*, London, Wiley, 1974.

19. BAUER, R. S. (Ed.), *Epoxy Resin Chemistry*, Amer. Chem. Soc. Symposium Series, **114**, 1979.
20. CRIVELLO, J. V., ref. 14, Chapter 2.
21. CRIVELLO, J. V., *Chem. Tech.*, 1980, p. 624.
22. PAPPAS, S. P. and JILEK, J. H., *Phot. Sci. Eng.*, **23** (1979), p. 140.
23. LEDWITH, A., *Polymer*, **19** (1978), p. 1217; *Makromol. Chem.*, Suppl. 3 (1979), p. 348.

Chapter 3

INTRA-MOLECULAR REACTION AND GELATION IN CONDENSATION OR RANDOM POLYMERISATION

R. F. T. Stepto

The University of Manchester Institute of Science and Technology, UK

SUMMARY

The occurrence of intra-molecular reaction in random (or condensation) polymerisations and associated fields of study are broadly reviewed. The principles of intra-molecular reaction, their application to ring-chain equilibria, and the interpretation of data using Jacobson–Stockmayer theory are discussed. Results on ring formation in a linear, sequential (or addition) polymerisation are presented as a link between ring-chain equilibria and intra-molecular reaction in irreversible, linear random polymerisations. The latter type of polymerisation is then considered. Data on total ring fractions are discussed in qualitative terms, and also interpreted using theories of irreversible, linear random polymerisations, which are themselves presented in a unified manner. Experimental results on pre-gel intra-molecular reaction in non-linear polymerisations are described and related to results from linear systems. An account is also given of the general gelation condition of Stockmayer and of the effects of intra-molecular reaction on the gel point. Finally, the interpretation of gelation data in terms of approximate gelation theories is considered, showing how intra-molecular reaction is determined by reactant structure and molar mass, as for linear polymerisations.

GLOSSARY OF SYMBOLS

b	effective bond length for a chain of n real, skeletal bonds such that $\langle r_n^2 \rangle = nb^2$
c_0, c_{a0}, c_{b0}	initial (molar) concentrations of reactive groups
c_a, c_b	(molar) concentrations of monomers or prepolymer reactants
c_a', c_b'	instantaneous concentrations of reactive groups
c_{ac}, c_{bc}	gel-point concentrations of reactive groups
$c_{ext}, c_{a,ext}, c_{b,ext}$	concentrations of reactive groups external to a given molecule
c_i	chain of i repeating units and concentration thereof
c_{int}	internal mutual (molar) concentration of pairs of reactive groups on the same molecule
$c_{int,i}$	c_{int} for a chain forming a ring or ring structure of i repeating units
C_{iA}, C_{iB}	concentrations of states of structural units (cascade theory)
$c_\omega, c_\alpha, c_\sigma$	concentrations of unreacted, inter-molecularly reacted and intra-molecularly reacted groups (cascade theory)
f, f_a, f_b	(chemical) functionalities of reactants
f_{aw}, f_{bw}	weight-average functionalities of reactants
i	number of repeating units in a chain forming a ring or ring structure
k_b	monomolecular rate constant for ring or chain scission
k_c	bimolecular rate constant for inter-molecular (chain) reaction
$k_{r,i}$	monomolecular rate constant for intra-molecular (ring) reaction to form a ring of i repeating units
K_i	molar, cyclisation equilibrium constant for rings of i repeating units in a linear polymerisation
M_a, M_b	molar masses of monomer or prepolymer reactants
M_n	number-average molar mass
n	number of skeletal bonds in a chain
N	Avogadro's number
N_a, N_b	numbers of monomers or prepolymer reactants in a polymerisation
N_r	number fraction of rings in a linear polymerisation or number of ring structures per molecule in a non-linear polymerisation

p, p_a, p_b	overall extents of reaction of reactive groups
p'	extent of reaction in chain fraction
p'_a, p'_b	instantaneous overall extents of reaction
Pab	mutual molar concentration of a pair of reactive groups which can react to form the smallest ring or ring structure
$P_{cA,i}, P_{cB,i}$	probabilities of occurrence of chains of i repeating units starting with an A group and a B group, respectively (cascade theory)
P_i	existence probability (unit fraction) of state of an A—A unit in an RA_2 polymerisation (rate theory)
P'_i	rate of change of P_i with respect to p
P_{iA}, P_{iB}	existence probabilities of states of A—A and B—B structural units in an $RA_2 + RB_2$ polymerisation (cascade theory)
P_{ij}	amount (unit fraction) converted from state i into state j of an A—A unit (rate theory)
P'_{ij}	rate of change of P_{ij} with respect to p
$P(\mathbf{0}, n)$	$P(\mathbf{r})$ at $\mathbf{r} = \mathbf{0}$ for a chain of n bonds
$P(\mathbf{r})$	probability density function of end-to-end vector ($\mathbf{r}$) of a linear chain
r	initial ratio of reactive groups (c_{a0}/c_{b0})
$\mathbf{r}$	end-to-end vector of a chain
r_i	instantaneous probability of forming a ring of i repeating units
$\langle r_n^2 \rangle, \langle r_{0,n}^2 \rangle$	mean-square end-to-end distances of chains and unperturbed chains of n bonds
R	total ring concentration
R_4	unit (weight) fraction of smallest rings in an RA_2 polymerisation (rate theory)
R_7	unit fraction of smallest rings in an $RA_2 + RB_2$ polymerisation (rate theory)
R_c	percentage of intra-molecular reaction at gelation
R_i	ring of i repeating units and (molar) concentration thereof
v_0	reaction volume for a pair of groups
α	$= p_a p_b$ in gelling systems
$\alpha, \alpha_a, \alpha_b$	extents of inter-molecular reaction in linear polymerisations
α_c	value of $p_a p_b$ at the gel-point
λ	ring-forming parameter (eqn. 58)

λ_{ab}	ring-forming parameter of Ahmad–Stepto expression for the gel point (eqns. 87 and 88)
λ'_{ab}	$= \lambda_{ab}/(1 - \lambda_{ab})$
λ_F	ring-forming parameter of Frisch gelation theory (eqn. 73)
λ_K	ring-forming parameter of Kilb gelation theory (eqn. 74)
v	number of bonds in the chain forming the smallest ring or ring structure in a polymerisation
$\sigma, \sigma_a, \sigma_b$	extents of intra-molecular reaction in linear polymerisations
$\sigma_{c,i}$	symmetry number of a chain of i repeating units
$\sigma_{r,i}$	symmetry number of a ring of i repeating units
$\phi(y, s)$	$= \sum_{i=1}^{\infty} y^i i^{-s}$
$\omega, \omega_a, \omega_b$	fractions of unreacted groups in linear polymerisations (cascade theory)

GLOSSARY OF CHEMICAL ABBREVIATIONS

ac	acid chloride
AC	adipoyl chloride
D	dimethylsiloxane unit ($-Si(CH_3)_2O-$) in a linear chain
DDI	decamethylene di-isocyanate
HDI	hexamethylene di-isocyanate
LG56	polyoxypropylene triol based on glycerol
LHT112	polyoxypropylene triol based on hexane-1,2,6-triol
LHT240	polyoxypropylene triol based on hexane-1,2,6-triol
PDMS	poly(dimethylsiloxane)
PEG	poly(ethylene oxide) diol (polyethylene glycol)
PEG200	polyethylene glycol, $M_n \simeq 200$
PEG400	polyethylene glycol, $M_n \simeq 400$
POE	polyoxyethylene
POP	polyoxypropylene
R	dimethylsiloxane unit in a ring
SC	sebacoyl chloride

1. INTRODUCTION

A classification of polymerisations has to be somewhat cumbersome if it has to relate to both the growth mechanism and the polymer formed. Flory[1a] and Lenz[2] have discussed the shortcomings of the original condensation/

addition classification of Carothers,[3] and Lenz has proposed that a proper classification should describe the growth mechanism, the relationship between monomer and repeating-unit structures, and the low or high molar mass of the polymer formed. However, from the point of view of polymerisation statistics, concerned principally with growth mechanisms and resulting molar mass and complexity distributions, such a classification is unnecessarily complicated. On statistical grounds, a basic classification into only two categories is needed, as in the original Carothers classification. Hence, the terms random polymerisation (as used in the title) and sequential polymerisation have been proposed[4] to describe the two main classes of polymerisation: that in which the basic mechanism is the reaction together of molecules of all sizes and that in which it is the addition of monomer units to growing chains. With regard to growth mechanism, the classification of course corresponds to the step-growth/chain growth classification of Lenz or what is normally meant by the condensation/ addition classification of Carothers, but has the advantages that the adjectives used are of more significance statistically and are devoid of inference concerning the chemistry of the reactions involved in the polymerisation.

In use, a basic classification always needs qualification to be precise in its description of the polymerisation under consideration. Thus, for example, an RA_2, equi-reactive, linear, random polymerisation leads to the most-probable distribution of molar masses, and an instantaneously-initiated, unterminated, sequential homopolymerisation leads to a Poisson distribution. Such qualifications to the basic classification may of course be introduced when required, and, in this context, the present chapter is concerned with the measurement and interpretation of intra-molecular reaction during equi-reactive, linear, reversible and irreversible, and equi-reactive, non-linear, irreversible, random polymerisations.

The classical works of Flory[1b,1c,5] and Stockmayer[6,7] relating to the polymerisation statistics of linear and non-linear random polymerisations still form the basis of present-day understanding of random polymerisation processes. The classical approach assumes that like groups are of equal reactivity, independent of the size of molecule to which they are attached, and that (before gelation) all reaction is inter-molecular. Generally later treatments have considered a wider range of polymerising systems, including the effects of like groups having intrinsic or induced[8] unequal reactivities (for example refs. 9–34) or have considered the effects of intra-molecular reaction (for example refs. 9, 19–22, 27, 33, 35–59), or both (refs. 9, 19–22, 27, 33).

The neglect of unequal reactivity in the classical treatments is a more obvious shortcoming than the neglect of intra-molecular reaction. It is, in principle, more easily treated, although detailed equations for the distributions of species or gelation conditions may be complicated.[10–13,28,34] Only the properties of pairs of groups about to react are required. Thus, groups can be effectively isolated from the molecules to which they are attached. In contrast, the probability of a pair of groups on the same molecule reacting together is intimately related to the length and structure of the chain connecting the two groups and (in non-linear polymerisations) to the structure of the rest of the molecule.

Unlike unequal reactivity, intra-molecular reaction is part of the very nature of the random polymerisation process. Random polymerisation can only continue if growing molecules contain at least two reactive groups, and there is always the chance that two groups from the same molecule can react together, provided that they are chemically capable of mutual reaction. The actual probability of intra-molecular reaction may vary, as may be deduced from Fig. 1, where c_{int} is the internal concentration of B groups around the reference A group and c_{ext} is the external concentration of B groups (from other molecules). The probability that a given A group reacts intra-molecularly rather than inter-molecularly depends on the concentration of B groups in its vicinity from the same molecule compared with the

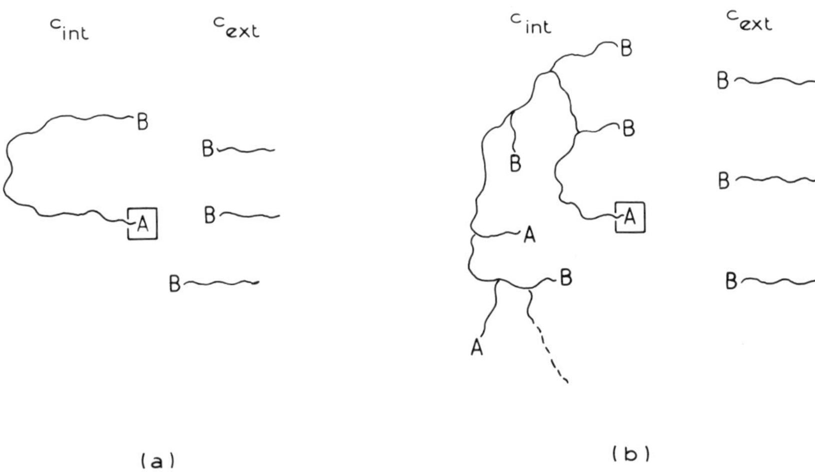

FIG. 1.   Concentrations of B groups around a reference $\boxed{A}$ group in (a) an $RA_2 + RB_2$ polymerisation and (b) an $RA_2 + RB_3$ polymerisation; c_{int} is the concentration of B groups from the same molecule, and c_{ext} that of groups from other molecules.

concentration of B groups from other molecules. The chain length between the A group and the nearest possible B group capable of intra-molecular reaction is defined by the structures of the monomer or prepolymer reactants. If it is sufficiently large, intra-molecular reaction may be neglected, irrespective of the number of B groups on the same molecule.

Intra-molecular reaction may sometimes be neglected in linear random polymerisations. In terms of Fig. 1(a) there can be only one pair of groups per molecule which can so react. In addition, polymerisations are often carried out in bulk when c_{ext} has its maximum values (as defined by the density, molar masses and extents of reaction of the reactants). However, in non-linear polymerisations the average number of reactive groups per molecule increases with extent of reaction and intra-molecular reaction cannot usually be neglected,[60] with important consequences. For example, gelation occurs at extents of reaction higher than those predicted by classical theory[8,48,50,59-62] and the properties of the materials formed post-gel are strongly influenced by the pre-gel intra-molecular reaction which has occurred.[63-65] The gel-point is of course of importance in both natural and industrial polymerisation processes as it defines the point of infinite viscosity in a liquid polymerising system.

With the occurrence of intra-molecular reaction, different distributions of molecular species result from polymerisations proceeding via irreversible and reversible reactions. In the former case, the c_{ext} in Fig. 1(a) and (b) decreases as reactions proceed, resulting in increases in the instantaneous probabilities of intra-molecular reaction for given species. In addition, ring structures, once formed, take no further part in the polymerisation processes. In contrast, polymerisations proceeding via reversible reactions can be made to yield ring-chain equilibriates, corresponding to fixed extents of reaction and single values of c_{ext}, and most studies of such polymerisations have involved ring-chain equilibriates of linear polymers, for example refs. 35, 40, 53, 54 and 57. Although many linear condensation polymers of commercial importance (e.g. nylons, polyesters) are prepared via chemically reversible reactions, studies of ring-chain equilibria may not have direct relevance to industrial polymerisations as the latter are often carried out 'irreversibly', with removal of the small molecular species formed. The study of intra-molecular reaction under such conditions is more complex than under equilibrium or truly irreversible conditions and has received little attention, either theoretically or experimentally, in the published literature.

In spite of the foregoing, experimental studies of intra-molecular reaction in linear random polymerisations do have important roles to play as the molecular species formed have much simpler structures than those formed in non-linear polymerisations, and the molecular populations are more amenable to analysis, both experimental and theoretical. Semlyen and collaborators, in particular,[54] have measured distributions of cyclic species in linear ring-chain equilibriates in bulk and in diluted systems, providing convincing support for the Jacobson–Stockmayer treatment of ring-chain equilibria[35] and for the rotational-isomeric state (R-I-S) model of polymer configuration.[66] In addition, Stepto and Waywell[44] have measured total ring fractions during irreversible linear polymerisations, providing data to test theories[44–46,51] of irreversible linear random polymerisation which include intra-molecular reaction. Further, Semlyen and collaborators have recently used ring-chain equilibriates of poly-(dimethylsiloxane) (PDMS) to prepare a unique series of macrocyclic species of up to several hundred atoms, enabling the properties of such molecules to be studied and compared with their linear counterparts.[67]

Intra-molecular reaction has been studied extensively in non-linear polymerisations. The gel-point can be determined relatively easily and related to intra-molecular reaction.[8,9,19,20,41,48,50,59–65,68–80] In addition, total ring fraction data have been obtained for irreversible polymerisations[60] and are currently being used in the author's laboratory to test theories of non-linear random polymerisation.[20,47,81]

Many of the gelling systems which have been studied have contained like groups of unequal reactivity, and, because the gel-point is a single point in a polymerisation which may be affected similarly by unequal reactivity and intra-molecular reaction, it is difficult to separate unequivocally the effects of the two phenomena. Many of the gelation data[9,19,20,41,68–72,76,77] come from such systems. However, this account will consider mainly data from systems where the assumption of equal reactivity is justified.[8,48,59–65] Of course, many gelling systems of industrial importance have like groups of unequal reactivity, and attempts may be made to account for the effects of intra-molecular reaction and substitution effects (induced unequal reactivity) on the gel-point and post-gel materials properties using cascade theory, as developed by Gordon, Dušek and collaborators,[19–22,33,47,55,56,58] and as described in Chapter 4.

2. PRINCIPLES OF INTRA-MOLECULAR REACTION IN RANDOM POLYMERISATIONS

As Fig. 1 shows, the competition between intra-molecular and inter-molecular reaction is the competition between pairs of groups which are spatially related (intra-molecular) and pairs of groups which are otherwise spatially independent (inter-molecular). Although the total number of ring structures formed as a polymerisation proceeds depends on the type of polymerisation occurring, the instantaneous factors affecting intra-molecular–inter-molecular competition are to a first approximation independent of the type of polymerisation.

The factors were first clearly defined by Jacobson and Stockmayer[35] in the context of linear random polymerisation. For each pair of A and B groups, say, which can react intra-molecularly (such as the terminal groups of A$\sim\sim$B molecules in an $RA_2 + RB_2$ random polymerisation)

$$c_{\text{int}} = \int_{v_0} P(\mathbf{r})d\mathbf{r} \Big/ \int_{v_0} d\mathbf{r} = P(\mathbf{0}) \cdot v_0/v_0 = P(\mathbf{0})$$

(functional groups per unit volume)   (1)

where $P(\mathbf{r})$ is the probability density function of end-to-end vectors $\mathbf{r}$ for the chain separating the pair of groups. According to eqn. 1, intra-molecular reaction is assumed to occur when the reacting end groups are within a volume v_0 of each other, sufficiently small for $P(\mathbf{r})$ to be taken as constant over v_0.

The simplest realistic expression that can be used for $P(\mathbf{0})$ in eqn. 1 is that for a Gaussian chain, giving for a chain of n bonds,

$$c_{\text{int},n} = P(\mathbf{0}, n) = \left(\frac{3}{2\pi\langle r_n^2 \rangle}\right)^{3/2} \cdot \frac{1}{N}$$

(mol functional groups per unit volume)   (2)

where N is Avogadro's number, and $\langle r_n^2 \rangle = nb^2$, with n the number of actual bonds in the chain forming the ring structure and b their effective bond length in the chain. Thus c_{int} decreases with chain stiffness (b) and with chain length (n). Equation 2 assumes that n is sufficiently large for Gaussian statistics to be obeyed, at least for $\mathbf{r} \cong \mathbf{0}$. In addition, because $P(\mathbf{r})$ is spherically symmetrical, it implies that any effects of chain-end orientation on the probability of reaction are the same for intra-molecular as for inter-molecular reaction.[40,53,54,57] Further, the application of eqn. 2 to non-linear polymerisations assumes that the various sub-chains, connecting pairs of groups which can react intra-molecularly, independently obey Gaussian statistics.[48]

In most random polymerisations only certain values of n can occur, as defined by the repeating unit of the chain structure. Two cases, giving the

FIG. 2.   Repeating units of the chain structures in (*a*) an $RA_2 + RB_2$ polymerisation and (*b*) an $RA_2 + RB_3$ polymerisation. The repeating units have v bonds separating the reacting groups, $\boxed{A}$ and $\boxed{B}$.

smallest values of n (those for repeating units) are illustrated in Fig. 2. In each case the smallest value of n is denoted by v. Thus, provided, principally, v is not too small, c_{int} for i repeating units ($i \times v$ bonds) can be written as

$$c_{int,i} = Pab/i^{3/2} \tag{3}$$

where

$$Pab = \left(\frac{3}{2\pi vb^2}\right) \cdot \frac{1}{N} \tag{4}$$

is characteristic of the monomer or prepolymer reactants.

These relations are similar to those used in the description of network formation in radical polymerisations (see Chapter 4). They have their origins in work by Kuhn,[82] and had been applied to radical polymerisations[83] prior to their application to random polymerisations.[35]

3.   LINEAR RING-CHAIN EQUILIBRIA

The incorporation of c_{int} into a description of a random polymerisation is most easily followed by considering the Jacobson–Stockmayer treatment[35] of linear ring-chain equilibria in an RA_2 polymerisation. The equilibria can be written as

$$C_{i+j} \rightleftharpoons C_j + R_i \qquad (\text{all } i > 0,\, j > 0) \tag{5}$$

where C_{i+j} and C_j denote chains of $i+j$ and j repeating units and R_i denotes a ring of i repeating units. For a given i and j, the equilibrium may be split into the two constituent equilibria

$$C_i + C_j \underset{k_b}{\overset{k_c}{\rightleftharpoons}} C_{i+j} \tag{6}$$

$$C_i \underset{k_b}{\overset{k_{r,i}}{\rightleftharpoons}} R_i \tag{7}$$

Assuming bimolecular chain reactions, the rates of the forward and backward reactions in each of eqns. 6 and 7 may be equated, giving at equilibrium

$$\text{rate}_{f,c} = k_c C_i \sigma_{c,i} \cdot C_j \sigma_{c,i} = \text{rate}_{b,c} = k_b C_{i+j} \sigma_{c,i+j} \tag{8}$$

and

$$\text{rate}_{f,r} = k_{r,i} C_i \sigma_{c,i} = \text{rate}_{b,r} = k_b R_i \sigma_{r,i} \tag{9}$$

respectively. Here, the symbols C_i, C_j, and R_i have also been used to represent the molar concentrations of the species they denote. $\sigma_{c,i}$, $\sigma_{c,j}$, and $\sigma_{c,i+j}$ are the symmetry numbers of the chain species involved in the reactions, and $\sigma_{r,i}$ is the symmetry number of a ring of i repeating units. The symmetry numbers of the chain species are all equal to 2 and $\sigma_{r,i} = 2i$. For the forward reactions, the symmetry numbers allow concentrations to be expressed in terms of reactive groups, that is chain ends. For the backward (scission) reactions the symmetry numbers denote the numbers of equivalent ways scission may occur.

The equilibrium constant for eqn. 5 in terms of concentrations of reactive groups is

$$K'_{i,j} = \frac{C_j \cdot R_i \sigma_{r,i}}{C_{i+j}} \tag{10}$$

Equation 8 gives

$$\frac{k_b}{k_c} = \frac{C_i \sigma_{c,i} \cdot C_j}{C_{i+j}} \tag{11}$$

and eqn. 9 allows $C_i \sigma_{c,i}$ to be eliminated from this equation, so that

$$\frac{k_{r,i}}{k_c} = \frac{R_i \sigma_{r,i} \cdot C_j}{C_{i+j}} = K'_{i,j} \tag{12}$$

enabling $K'_{i,j}$ to be expressed as a ratio of the rate constants for ring and chain reaction. Note that k_b has been eliminated in going from eqn. 11 to eqn. 12 as no distinction was drawn between the k_b in eqn. 6 and that in eqn. 7. Thus, the ease with which a bond can break is assumed independent of whether that bond is in a ring or a chain.

In terms of molar concentrations of species, eqn. 12 may be written as

$$K_{i,j} = \frac{k_{r,i}}{k_c \sigma_{r,i}} = \frac{R_i \cdot C_j}{C_{i+j}} \tag{13}$$

Further, $K_{i,j}$ may be denoted K_i because C_j/C_{i+j} is independent of j, it being equal to $(p')^i$, where p' is the extent of reaction in the chain fraction.[35,40,54]

Alternatively p' may be defined as the ratio of the measured concentrations of chain species of degrees of polymerisation i and $i-1$.[40,54] Thus,

$$K_i = \frac{k_{r,i}}{k_c \sigma_{r,i}} = \frac{R_i}{(p')^i} \tag{14}$$

In experimental studies of ring-chain equilibriates, the quantities on the right-hand side of eqn. 14 are those measured experimentally and correlated with theoretical expressions for the ratio of rate constants, $k_{r,i}/k_c$. The rate constant $k_{r,i}$ is usually expressed as the product of k_c (which may be considered as the intrinsic rate constant for the reaction of pairs of groups) and the mutual concentration of chain ends, $c_{int,i}$, given by eqn. 3. Thus, substituting for $\sigma_{r,i}$, the Jacobson–Stockmayer expression[35] for the concentration of i-mer rings results, namely,

$$R_i = \left(\frac{3}{2\pi v b^2}\right)^{3/2} \cdot \frac{1}{i^{5/2} 2N(p')^i} = \frac{Pab}{i^{5/2} 2(p')^i} \tag{15}$$

The present derivation differs from others in the literature[35,40] in that it does not invoke the representation of K_i in thermodynamic terms. Similar expressions result for R_i in ARB and $RA_2 + RB_2$ polymerisations.[35] They differ only in the value of the symmetry number $\sigma_{r,i}$ and in the factor $(p')^i$.

It is to be expected that eqn. 15 will hold only over a certain range of values of $n\ (=iv)$. For rings which are too small, the assumption of a Gaussian density function, giving the factor $(3/2\pi\langle r_n^2 \rangle)^{3/2}$, will not be justified, nor its factorisation into $(3/2\pi v b^2)^{3/2}/i^{3/2}$. In general, b increases with chain length, but, typically, has increased for chains of about 30 bonds to within 5–10% of its limiting value for an infinite chain. However, the factorisation of $\langle r_n^2 \rangle$ to ivb^2 can in principle be avoided as realistic values of $\langle r_n^2 \rangle$ can be calculated using matrix methods,[66] provided the statistical weights of the various skeletal bond conformations are known for the chain structure under consideration. Thus eqn. 15 may be written:[40,53,54]

$$R_i = \left(\frac{3}{2\pi\langle r_n^2 \rangle}\right)^{3/2} \frac{1}{i2N(p')^i} \tag{16}$$

Also for small n, the mutual orientation of the groups at the ends of a chain with $\mathbf{r} \simeq 0$ will no longer be random, as it would be for two groups from different molecules. Thus, the probability that orientations occur which are suitable for reaction is different for intra-molecular as compared with inter-molecular reaction. Orientation effects have been considered by Flory and Semlyen and coworkers,[40,53,54,57] and the definition of c_{int} in eqn. 1 has to be modified by the introduction of a factor giving the probability of end-group orientations which are suitable for reaction. At very small n, only, rings can be formed which are subject to bond-angle and torsional strain. In such cases, not only would the expression for c_{int} need to be changed, but also the assumptions that k_b and k_c are equal for intra-molecular and inter-molecular reaction. Finally, for large n, deviations from eqn. 15 will be expected if the ring-chain equilibrium is carried out under non-theta conditions, when chain expansion will effectively increase b and reduce the concentration of cyclics. Equation 16 can be used to take approximate account of the effects of chain expansion, provided uniform expansion is assumed.

Experimental data on ring-chain equilibriates and their interpretation have recently been reviewed by Semlyen,[54] and the reader is referred to that text for detailed consideration of particular systems. In general, the concentrations and chain species in equilibriates have been determined using gas–liquid chromatography (g.l.c.) and gel permeation chromatography (g.p.c.). Data on a wide variety of systems now exist and in most cases satisfactory quantitative explanations of the data have been achieved on the bases of eqns. 15 and 16. As an example, plots of log K_i versus log i for PDMS equilibriates are shown in Fig. 3. The agreement between theory and experiment is good for $30 < n < 400$ (i.e. $15 < i < 200$), with calculated K_i values lying within 10–20 % of the experimental ones. The deviations at smaller n are due to the factors previously discussed and have received semi-quantitative explanations from Suter $et\ al.$[53b]

A detailed investigation of K_i values for PDMS in bulk, in diglyme (a poor solvent), and in toluene (a good solvent) have been carried out by Wright.[85] The results for the larger rings are shown in Fig. 4. The increasing deviations in the K_i values for toluene below the calculated curve as i increases are consistent with the effects of chain expansion. Note that the solutions do not refer to infinite dilution, but to approximately 20 % concentration, so only reduced effects of chain expansion will be observed. Even so, it is interesting to note that the curve for the toluene system starts to deviate noticeably from the other two experimental curves at about 100 bonds. This is the same region in which excluded volume effects are first apparent from measurements of diffusion coefficients of linear and cyclic PDMS in toluene at 298 K.[67e]

The general picture with regard to ring-chain equilibria in linear random polymerisations is that the Jacobson–Stockmayer expression, modified as in eqn. 16, provides an adequate prediction of the ring population provided the rings are not too small. Further, provided the effects of chain expansion

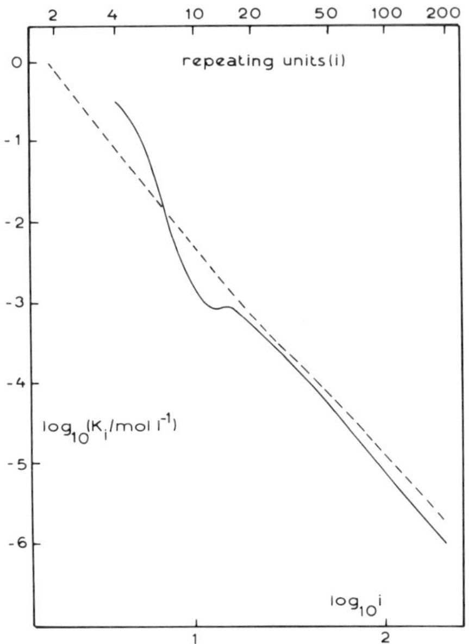

FIG. 3. Molar cyclisation equilibrium constant (K_i) versus ring size (i) for PDMS equilibriates in toluene.[40] ———, data for 383 K from Brown and Slusarczuk;[84] ----, values of log K_i calculated[40] from eqns. 14 and 16 using the Flory–Crescenzi–Mark R-I-S model for unperturbed dimensions $\langle r^2_{0,n} \rangle$, with $n = 2i$, at 383 K. (Reproduced in part from Flory, P. J. and Semlyen, J. A., *J. Amer. Chem. Soc.*, **88** (1966), p.3209, by permission of American Chemical Society Ⓒ.)

are not too large, values of K_i can be calculated independently from values of $\langle r^2_{0,n} \rangle$, thus furnishing important justification of the R-I-S model of polymer chain configuration and the statistical weights assigned for various chain structures.[54,66]

From eqn. 15 the predictions are:

(i)   The *molar concentrations* of rings are independent of the dilution of the reaction mixture provided the configurational properties of the chains remain unaltered. Thus, for example, when preparing ring-chain equilibriates, relatively more rings are formed in dilute solutions as the molar concentrations of chain species of course decrease with dilution.

(ii) R_i is given by an expression of the form

$$R_i = \kappa \cdot i^{-5/2} \tag{17}$$

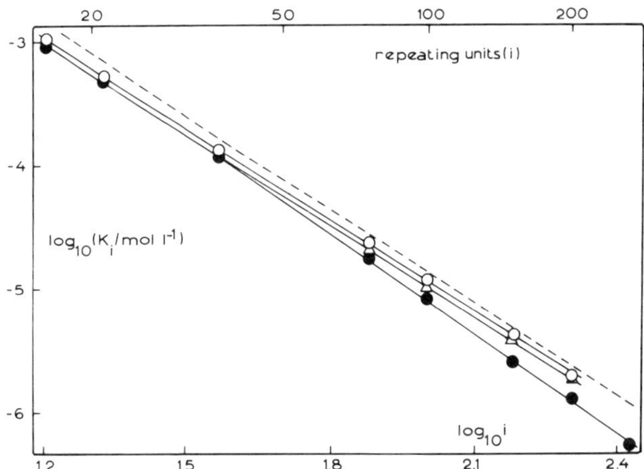

FIG. 4. Molar cyclisation equilibrium constant (K_i) versus ring size (i) for PDMS equilibriates in bulk at 383 K ($\bigcirc$), at 212 g litre^{-1} in diglyme at 333 K ($\triangle$) at 224 g litre^{-1} in toluene at 383 K ($\bullet$).[85] – – – log K_i calculated as in Fig. 3. (Reproduced from Wright, P. V., *J. Polym. Sci., Polymer Phys. Ed.*, **11** (1973), p. 51, by permission of John Wiley and Sons, Inc. ©.)

The factor $i^{-5/2}$ may be considered as the product of $i^{-3/2}$ and i^{-1}. The former relates to the probability that a chain of i repeating units has its ends coincident and the latter to the number of points at which a ring of i units may break.

(iii) The constant of proportionality κ in eqn. 17 decreases with increasing chain stiffness (b) and repeating-unit chain length (v).

It is well known that the most-probable distribution for ring-free polymerisations can be derived using (bimolecular) kinetics or statistical arguments,[86,87] independent of the actual kinetics of the chemical reactions involved. However, it is important that if the detailed expressions for rates of reaction contain additional factors, due to catalysis, etc., then such factors should be present for both the inter-molecular and intra-molecular reactions. In the context of eqns. 8 and 9, such factors should also be independent of i and j so that the random reaction between species, subject to the conditions for intra-molecular reaction, is preserved. Some discussion of these points with reference to polyurethane formation has been given previously.[51,59]

4.   CYCLISATION IN A SEQUENTIAL POLYMERISATION

Chojnowski *et al.*[88] have studied the irreversible formation of cyclic oligomers produced in the first step of the cationic polymerisation of hexamethylcyclotrisiloxane (R_3) and the results form an interesting comparison with those from ring-chain equilibria in PDMS.

The active chains formed from the cyclic monomer may be written $B(D_3)_q A$, where the groups A and B depend on the catalyst used and D denotes the dimethylsiloxane unit, $—Si(CH_3)_2O—$. The growth mechanism proposed is

$$\longrightarrow B(D_3)_q A \xrightarrow[k_p]{D_3} B(D_3)_{q+1} A \xrightarrow[k_p]{D_3} \cdots$$

$$\downarrow k_{r,q} \qquad\qquad \downarrow k_{r,q+1}$$

$$R_{3q} \qquad\qquad\quad R_{3(q+1)}$$

where R_{3q} and $R_{3(q+1)}$, etc., are cyclic species. Using the same symbols for molar concentrations, their rates of formation are

$$\frac{dR_{3q}}{dt} = k_{r,q} \cdot B(D_3)_q A \tag{18}$$

$$\frac{dR_{3(q+1)}}{dt} = k_{r,q+1} \cdot B(D_3)_{q+1} A \tag{19}$$

and so on. Provided chain propagation is much faster than cyclisation ($k_p D_3 \ll k_{r,q}$) then the concentrations of the intermediates, $B(D_3)_q A$, $B(D_3)_{q+1} A$, etc., may be assumed equal for q much less than the average degree of polymerisation (d.p.) of the polymer formed. In this case, the relative concentrations of cyclic species are independent of the concentrations of their respective linear intermediates, with

$$\frac{dR_{3q}}{dR_{3(q+1)}} = \frac{k_{r,3q}}{k_{r,3(q+1)}} = \frac{(R_{3q})_t}{(R_{3(q+1)})_t} \tag{20}$$

where $(R_{3q})_t$ is the value of R_{3q}, at time t. As previously, $k_{r,3q} = k_c \cdot c_{int,3q}$, giving in the Gaussian approximation (cf. eqns. 3 and 4)

$$R_i = \kappa' i^{-3/2} \tag{21}$$

with $i = 3q, 3(q+1) \ldots$. The form of this dependence of R_i on i may be contrasted with that of eqn. 17 for ring-chain equilibrium. Also in eqn. 17, the constant of proportionality is defined explicitly, whereas in eqn. 21 direct integration of eqn. 18 or 19 shows that κ' depends on t and the values assumed for k_c and the concentration of linear intermediates.

The concentrations of cyclic species from polymerisations of R_3 in various solvents with various catalysts were analysed by g.l.c. The concentrations of the smaller rings depended on the catalyst and solvent used, indicating a dependence on the detailed reaction mechanism and a sensitivity to chain-end orientation. The dependence reduced as the ring size increased and for catalysts which gave exclusively the cyclic species R_{3q}, $R_{3(q+1)}\ldots$, it was found that eqn. 21 was obeyed for $3q \geq 18$. Typical results are shown in Fig. 5 and compared with those from a PDMS ring-chain equilibriate.

The important points of comparison are the similarity in shapes of the two curves and their slopes at the larger values of i, rather than the absolute values of $\log R_i$. For the equilibriate, the slope agrees approximately with that predicted by Jacobson–Stockmayer theory and for the kinetically

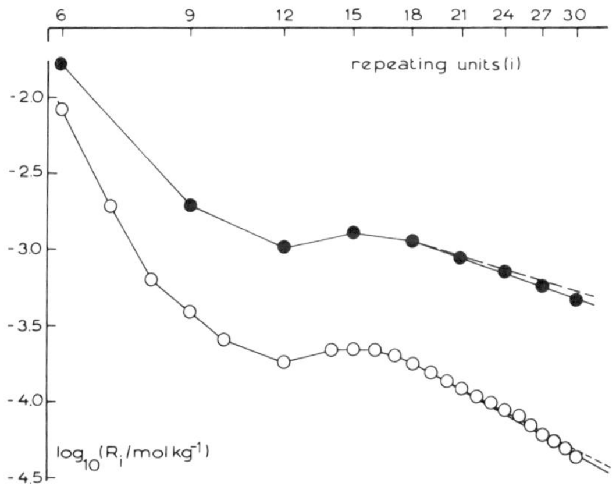

FIG. 5. Comparison of kinetically controlled distribution of cyclic oligomers in the polymerisation of hexamethylcyclotrisiloxane (40 % w/w solution in heptane at 30 °C) with CF_3SO_3H (7×10^{-4} mol kg^{-1}), at 30 % monomer conversion ($\bullet$) with the distribution of polydimethylcyclosiloxanes in equilibrium with the chain polymer in 40 % w/w solution ($\bigcirc$). Equilibrium concentrations (in mol kg^{-1}) of oligomers were multiplied by 0·3 to make them comparable with the concentrations controlled by kinetics;[88] ---, lines of slopes $-3/2$ and $-5/2$, in accordance with eqns. 21 and 17, respectively. (Reproduced from Chojnowski, J., Ścibiorek, M. and Kowalski, J., *Makromol. Chem.*, **178** (1977), p. 1351, by permission of Hüthig and Wepf Verlag, Basel ©.)

controlled case, the slope is approximately equal to $-3/2$, the value predicted for irreversible reaction with *equal* concentrations of the corresponding linear species. The similarity of the shapes of the two curves at smaller i values indicates that the deviations from linearity have common causes which depend on chain configurational properties rather than polymerisation type or reaction mechanism.

5.　INTRA-MOLECULAR REACTION IN IRREVERSIBLE LINEAR RANDOM POLYMERISATIONS

Irreversible linear random polymerisations including intra-molecular reaction are more difficult to treat both theoretically and experimentally than ring-chain equilibriates, as in the latter attention can be focused on one (equilibrium) extent of reaction (characterised by p'). For irreversible polymerisations it has not been possible to calculate ring concentrations *a priori*. Rather, effective values of the parameter Pab have been found which allow systematic interpretations of the data. In addition, the only available data on concentrations of ring species are those due to Stepto and Waywell[44] referring to polyurethane formation, and they give only total ring fraction data in contrast to the detailed data on concentrations of individual species available for equilibriates.

Comparison with the equilibrium case can be made by considering the development of the equations for an RA_2 polymerisation, corresponding to those in section 3 for ring-chain equilibrium. The growth reactions are

$$C_i + C_j \xrightarrow{k_c} C_{i+j} \qquad (i>0, j>0) \tag{22}$$

and

$$C_i \xrightarrow{k_{r,i}} R_i \tag{23}$$

giving for the rate of inter-molecular reaction of species i

$$\text{rate}_{c,i} = k_c C_i \sigma_{c,i} \sum_{j=1}^{\infty} C_j \sigma_{c,j} \tag{24}$$

and for its rate of intra-molecular reaction

$$\text{rate}_{r,i} = k_{r,i} C_i \sigma_{c,i} \tag{25}$$

In contrast to the equilibrium case, the sum of eqns. 24 and 25 gives the rate of *disappearance* of *i*-mer which can be combined with similar expressions for its rate of appearance to give the overall rate of change; compare:[87]

$$\frac{dC_i}{dt} = k_c \sum_{j=1}^{j<i/2} C_j \sigma_{c,j} \cdot C_{j-i} \sigma_{c,j-i} \qquad \text{(for } i > 1\text{)}$$

$$+ \tfrac{1}{2} k_c (C_{i/2} \sigma_{c,i/2})^2 \qquad \text{(for } i \text{ even)}$$

$$- k_c C_i \sigma_{c,i} \sum_{j=1}^{\infty} C_j \sigma_{c,j} - k_c (\text{Pab} \cdot i^{-3/2}) \cdot C_i \sigma_{c,i} \qquad (26)$$

$$\frac{dR_i}{dt} = k_c (\text{Pab} \cdot i^{-3/2}) C_i \sigma_{c,i} \qquad (27)$$

with $i = 1, 2 \ldots$. Here, use of the relations $k_{r,i} = k_c c_{\text{int},i} = k_c \text{Pab} \cdot i^{-3/2}$ has been made. Corresponding sets of equations can be written down for other types of linear random polymerisation. However, they all have to be solved numerically; see for example ref. 45.

As stated, only total ring-fraction data need to be considered at present, and for this purpose simplifications to the equations which need to be solved can be made. However, before discussing quantitative interpretations of such data, it is illustrative to view the data qualitatively. For this purpose, the *instantaneous* probability of formation of rings of a given size, r_i, is a useful parameter. To evaluate r_i, consider the summation in eqn. 24. This represents the instantaneous concentration of reactive groups in the reaction mixture and is analogous to the c_{ext} of Fig. 1. Thus,

$$r_i = \frac{\text{rate}_{r,i}}{\text{rate}_{r,i} + \text{rate}_{c,i}} = \frac{c_{\text{int},i}}{c_{\text{int},i} + c_{\text{ext}}} \qquad (28)$$

Note that eqns. 26 and 27 could be rewritten using r_i and c_{ext}, and that r_i increases as a polymerisation proceeds and c_{ext} decreases.

5.1. Total Ring Fractions in RA$_2$ + RB$_2$ Polymerisations

The polymerisations studied[44] were those forming linear polyurethanes from poly(ethylene oxide) diols (PEG) and hexamethylene diisocyanate (HDI). Reactions were in the absence of added catalyst and were carried out at 70 °C in benzene solution and in bulk. The number (mol) fraction of rings (N_r) was determined as a function of extents of reaction, using reaction mixtures at different dilutions and different ratios of reactants. Also, two diols having different molar masses were used.

The number fraction (N_r) was evaluated from a combination of measurements of extent of reaction and of the number average molar mass (M_n). The principle of the evaluation depends on the fact that every time a pair of groups reacts inter-molecularly, a molecule is lost to the system, whereas intra-molecular reaction leaves the number of molecules unchanged. Thus, by measuring, by titration, the number of pairs of groups reacted (extent of reaction) and the number of molecules present, the amount of intra-molecular reactions can be deduced. The general equation for determining N_r in an $RA_{f_a} + RB_{f_b}$ polymerisation is[60]

$$N_r = 1 - \frac{(1 + f_a/(f_b \cdot r) - f_a p_a)M_n}{M_a + f_a M_b/(f_b \cdot r)} \tag{29}$$

where M_a and M_b are the molar masses of the monomer or prepolymer reactants, p_a is the extent of reaction of titrated A groups and $r\,(= c_{a0}/c_{b0})$ is the initial ratio of reactive groups.

M_n was determined cryoscopically using benzene as solvent and an apparatus developed especially for the investigation[89] and the reasons for choosing the particular chemical reactants and methods of analysis have been discussed.[44] Firstly, no small molecule is formed in the $-OH +$ NCO— reaction. If small molecules were formed when pairs of groups reacted, then a correction to M_n would be required to discount their presence. Such a correction would reduce the accuracy of the method. Secondly, the temperature of the cryoscopic analysis was such that reaction during analysis could be neglected. Thirdly, kinetics studies[8,90,91] had shown that like functional groups of reactants of the types used have equal reactivities, resulting in fewer parameters in any theoretical interpretation of the results. Lastly, the values of v for the two reaction systems used were equal to 25·2 (HDI/PEG200) and 35·7 (HDI/PEG400), thus allowing the use of the previously discussed Gaussian-based expressions for $c_{\text{int},i}$.

The details of the reactions studied are given in Table 1, and plots of N_r as a function of reaction, p, are shown in Figs. 6 and 7. Figure 6 refers to experiments for which $r = 1$, and Fig. 7 to experiments carried out at 5% w/w monomers with different ratios of reactants. For $r \neq 1$, p refers to the extent of reaction of the minority group.

With reference to the results in Fig. 6, it can be seen that the curves of N_r versus p have the expected shapes and relative positions: N_r increases from zero at $p = 0$, tending to 1 at complete reaction and the *rate* of ring formation (slope) increases with p. The latter behaviour may be attributed

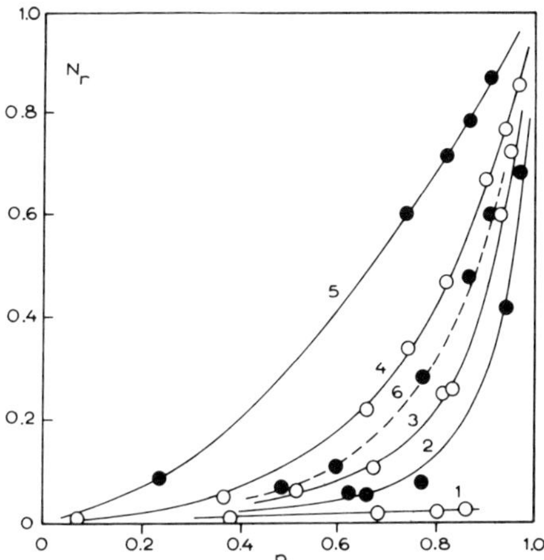

FIG. 6. Number fraction of rings (N_r) versus extent of reaction (p) for equimolar reaction mixtures ($r = 1$). Experiment numbers of Table 1 are given with the curves:[44] ——, HDI/PEG200; – – –, HDI/PEG400. (Reproduced from Stepto, R. F. T. and Waywell, D. R., *Makromol. Chem.*, **152** (1972), p. 263, by permission of Hüthig and Wepf Verlag, Basel ©.)

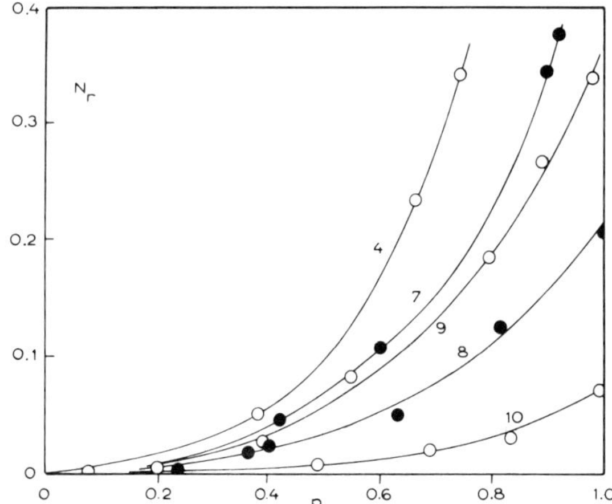

FIG. 7. Number fraction of rings (N_r) versus extent of reaction of minority group (p) for $r \neq 1$ and for experiment 4 of Table 1. All experiments for HDI/PEG200 at 5% w/w monomers.[44] (Reproduced from Stepto, R. F. T. and Waywell, D. R., *Makromol. Chem.*, **152** (1972), p. 263, by permission of Hüthig and Wepf Verlag, Basel ©.)

TABLE 1

INITIAL CONCENTRATIONS AND RATIOS OF REACTANTS USED IN THE STUDY[44] OF INTRA-MOLECULAR REACTIONS DURING LINEAR POLYURETHANE FORMATION

Experiment	$[NCO]_0$ $(mol\,kg^{-1})$	$[OH]_0$ $(mol\,kg^{-1})$	r^a	Percentage of monomers (w/w)
1	5·111	5·188	1	100 (bulk reaction)
2	0·9591	0·9777	1	20
3	0·5283	0·5333	1	10
4	0·2479	0·2483	1	5
5	0·0518	0·0525	1	1
6^b	0·2458	0·2457	1	5
7	0·3264	0·2093	1/0·64	5
8	0·4180	0·1382	1/0·33	5
9	0·1873	0·3124	0·60/1	5
10	0·1377	0·4156	0·33/1	5

^{a} $r = [NCO]_0/[OH]_0$.
^{b} HDI/PEG400, other experiments HDI/PEG200.
(Reproduced from Stepto, R. F. T. and Waywell, D. R., *Makromol. Chem.*, **152** (1972), p. 263, by permission of Hüthig and Wepf, Verlag, Basel ©.)

to the decrease in c_{ext} and resulting increase in r_i as the reaction proceeds. The relative positions of the curves show that ring formation increases with initial dilution (decrease of c_{ext}), and that an increase in v (from 25·2 to 35·7), giving a decrease in $c_{int,i}$ at the same initial concentrations of reactive groups (experiments 4 and 6), decreases ring formation.

The curves in Fig. 7 show broadly the expected variation in N_r with r, in which ring formation decreases as r moves away from unity, there being, for $r = 1$, a maximum fraction of the molecules (A⤳B) which can undergo intra-molecular reaction. In addition, curves for reciprocally related values of r should be coincident. The curves for experiments 7 and 9 are approximately in keeping with this expectation. (The r values of the two experiments are not exactly reciprocally related, with that for experiment 7 being nearer to unity. However, the initial concentration of reactive groups for experiment 9 is slightly less than for experiment 7.) Those for experiments 8 and 10 do show marked deviations from coincidence which may be attributed[44] to the occurrence of side reactions when there is a large excess of isocyanate groups (experiment 8).

Notice that for the bulk system (experiment 1) there is a negligible amount of ring formation. The values of N_r determined are essentially zero to within experimental error. This conclusion is valid for the system considered (HDI/PEG200), as characterised by its value of Pab $(=(3/2\pi vb^2)^{3/2}/N)$. For prepolymers or monomers having lower molar masses and more flexible chain structures, larger values of N_r may be expected, although they are unlikely to exceed 10% even at high conversions. This situation is in marked contrast to that for non-linear, $RA_2 + RB_3$ polymerisations (see section 7).

6. CORRELATIONS OF TOTAL RING FRACTION DATA IN IRREVERSIBLE LINEAR RANDOM POLYMERISATIONS

There have been four attempts[44-46,51] at the quantitative interpretation of the data presented in the preceding section, all of which have been able to fit the N_r versus p curves with some success, essentially in terms of the independent parameter b, as contained in Pab. Stepto and Waywell[44] used Jacobson–Stockmayer theory approximately modified for irreversibility, Gordon and Temple[45] analysed the results using simultaneous kinetics equations, and followed this with an analysis using cascade theory[46] and finally, Stanford et al.[51] used a limiting form of the rate theory[4,51] of random polymerisations. All of the approaches contain approximations, but the last two in particular can be extended to treat intra-molecular reaction in a wide variety of random polymerisations and will be discussed in more detail (sections 6.1 and 6.2).

Stepto and Waywell[44] assumed that Jacobson–Stockmayer theory applied, except that the concentration of rings of i repeating units should vary as $i^{-3/2}$ rather than $i^{-5/2}$ (see eqns. 21 and 17). This modification is the analogy for linear polymerisations of Kilb's modification[38] for the prediction of gel-points in non-linear polymerisations, and, as comparison with the work of Chojnowski et al.[88] shows, corresponds to the assumption of equal concentrations of chain species of all lengths. Complete equations have been published[35] only for $r = 1$, when the changed expression for R_i gives

$$\frac{R}{C_a + C_b} = \frac{N_r}{M_n} \cdot \frac{(M_a C_a + M_b C_b)}{(C_a + C_b)} = \frac{Pab \cdot \phi(p'^2, 3/2)}{2(C_a + C_b)} \tag{30}$$

$R = \Sigma R_i$, C_a and C_b are the (initial) molar concentrations of monomers, and

$$\phi(y, s) = \sum_{i=1}^{\infty} y^i i^{-s} \tag{31}$$

104   R. F. T. STEPTO

Thus, $\phi(p'^2, 3/2)$ arises from the summation over all ring sizes, required for the evaluation of R from R_i. In addition, the overall extent of reaction and that in the chain fraction are related by the equation

$$p = p' + 2(1 - p') \cdot \frac{\mathrm{Pab}}{2(C_a + C_b)} \cdot \phi(p'^2, 1/2) \qquad (32)$$

From eqns. 30 and 32, $R/(C_a + C_b)$ was plotted as a function of p for given values of Pab and compared with the experimental data plotted in the same form, allowing the values of Pab which give the closest agreement with experiment to be deduced. Equation 4 then allowed the best value of b, strictly the effective bond length of the chain forming the smallest ring, to be deduced.

The kinetics analysis of Gordon and Temple[45] used the set of simultaneous equations for an $RA_2 + RB_2$ polymerisation corresponding to eqns. 26 and 27 for the RA_2 case. The equations have t as independent variable; t is only true time if k_c and Pab are known and second-order kinetics observed. Also, the data concerned have the more fundamental quantity p_a or p_b as independent variable. Both these points require that solutions to the equations be presented as independent of t. This is easily achieved as direct solution of the kinetics equations gives the concentrations of both chain and ring species. These may be converted into α_a and σ_a, say, the fractions of A groups which have reacted inter-molecularly and intra-molecularly at a given value of $k_c t$. Further,

$$\alpha_a + \sigma_a = p_a \qquad (33)$$

and for $r = 1$

$$N_r = \sigma/(1 - \alpha) \qquad (34)$$

with corresponding expressions existing for $r \neq 1$. Thus eqns. 33 and 34 enable N_r to be evaluated as a function of p. The true independent parameter in the kinetics equations (as in eqns. 26 and 27) is Pab. By allowing the sets of equations to extend up to a d.p. of 120, good fits to the experimental N_r versus p curves were obtained and the best values of Pab and, hence, b determined.

6.1. Cascade Theory

The application of cascade theory to polymerisation processes was first described by Gordon[92] and since then its use has grown widely, e.g. as described in refs. 16, 19–22, 24, 26, 31–33, 43, 46, 47, 49, 50, 55, 56, 58, 93–100. To date, in applications to ring formation, only relatively simple considerations of cascade theory have been required and the present description will be limited to these.

The equations derived from cascade theory are usually presented in terms of link probability-generating-functions and their derivatives, e.g. in refs. 22, 46. However, such a representation is not essential and the expressions used by Gordon and Temple[46] to analyse the data of Stepto and Waywell may be derived and presented in a way more related to the preceding kinetics analysis, and without the reader needing to be acquainted with cascade formalism.

The equations relate to the states of the structural units involved in a polymerisation, the states being described in terms of the states of reaction of the functional groups. For an $RA_2 + RB_2$ random polymerisation, the states may be defined as in Table 2, with references to RA_2 units. In state 1 both A groups are unreacted, in 2 one has reacted inter-molecularly, in 3 both have reacted inter-molecularly, and in 4 the A group to the left has reacted inter-molecularly followed by the A group to the right reacting intra-molecularly with the B group at the end of the chain connected to the left-hand A group. According to the spanning-tree approximation,[46] the details of the ring structure in state 4 are not specified, and a summation is performed over the probabilities of all possible ring sizes, to give the total probability of intra-molecular reaction of the second (right-hand) A group of state 4.

The following kinetics rate equations are set up describing the interconversion of states. They are written in terms of concentrations but may equally well be written in terms of the existence probabilities P_{iA}.

$$\frac{dC_{1A}}{dt} = -k_c \cdot 2C_{1A} \cdot c_{b,ext} \atop (1A \rightarrow 2A)$$

$$\frac{dC_{2A}}{dt} = k_c \cdot \underset{(1A \rightarrow 2A)}{2C_{1A} \cdot c_{b,ext}} - k_c \cdot \underset{(2A \rightarrow 3A)}{C_{2A} \cdot c_{b,ext}} - \sum_{i=1}^{\infty} \underset{(2A \rightarrow 4A)}{k_{r,i} \cdot C_{2A} \cdot P_{CA,i}}$$

$$\frac{dC_{3A}}{dt} = k_c \cdot \underset{(2A \rightarrow 3A)}{C_{2A} \cdot c_{b,ext}}$$

$$\frac{dC_{4A}}{dt} = \sum_{i=1}^{\infty} \underset{(2A \rightarrow 4A)}{k_{r,i} \cdot C_{2A} \cdot P_{CA,i}} \tag{35}$$

R. F. T. STEPTO

TABLE 2

CASCADE THEORY:[46] STATES OF A—A (OR RA_2) STRUCTURAL UNITS IN AN $RA_2 + RB_2$ RANDOM POLYMERISATION[a]

State		Existence probability
1A	A—A	$P_{1A} = C_{1A}/C_A$
2A	—BA—A	$P_{2A} = C_{2A}/C_A$
3A	—BA—AB—	$P_{3A} = C_{3A}/C_A$
4A	⌐BA—AB	$P_{4A} = C_{4A}/C_A$

[a] A corresponding set exists for the B—B units. C_{iA} is the (molar) concentration of state iA and C_A the concentration of A—A units or monomers.

Each term on the right-hand side of eqn. 35 gives the rate of a particular interconversion reaction between states as specified under the individual terms, and the total rate of change $\sum_i dC_{iA}/dt = 0$, consistent with a constant overall concentration of A—A units. The variable $c_{b,ext}$ is the concentration of B groups external to the reference RA_2 unit (cf. Fig. 1). When considering the corresponding set of equations for RB_2 units, a factor $c_{a,ext}$ will occur. This can be defined in terms of the C_{iA}, namely,

$$c_{a,ext} = 2C_{1A} + C_{2A} \tag{36}$$

Thus, by symmetry,

$$c_{b,ext} = 2C_{1B} + C_{2B} \tag{37}$$

With respect to intra-molecular reaction $(2A \rightarrow 4A)$, $k_{r,i}$ is again the intra-molecular rate constant for forming a ring of i repeating units, and $P_{cA,i}$ is the probability of occurrence of a chain of i units. Thus, $k_{r,i} = k_c \cdot Pab \cdot i^{-3/2}$, and $P_{cA,i}$ is expressed in terms of the concentrations of the states of the A—A and B—B units. Consider $P_{cA,1}$; this is the probability that state 2A gives the species

B—B<u>A</u>—A

(where the reference unit is underlined) or, with respect to the set of B—B states, the probability that

B^2—B $^1A^1$—

occurs *given* that the A group (A^1) forms part of state 2A. Thus, the probability required is that a chosen B group (B^1) which has reacted *inter-molecularly* is part of a unit containing an unreacted B group (B^2). By symmetry from Table 2, the total concentration of inter-molecularly reacted B groups is $C_{2B} + 2C_{3B} + C_{4B}$. Hence,

$$P_{cA,1} = C_{2B}/(C_{2B} + 2C_{3B} + C_{4B}) \qquad (38)$$

The next ring forms from the chain

$$B\!-\!BA\!-\!AB\!-\!B\underline{A}\!-\!A$$

containing two repeating units. Moving left from the reference unit, the probabilities that $-AB\!-\!\underline{BA}-$, $-\underline{BA}\!-\!AB-$, and $\underline{B}\!-\!\underline{BA}-$ occur are required. The last probability is equal to $P_{cA,1}$, and, from reasoning similar to that leading to eqn. 38, the first two probabilities are

$$2C_{3B}/(C_{2B} + 2C_{3B} + C_{4B})$$

and

$$2C_{3A}/(C_{2A} + 2C_{3A} + C_{4A})$$

respectively. (Here, the numerator, $2C_{3B}$, is the concentration of inter-molecularly reacted B groups on units with a second inter-molecularly reacted B group, and $2C_{3A}$ is correspondingly interpreted.) Thus,

$$P_{cA,2} = \frac{2C_{3B}}{(C_{2B} + 2C_{3B} + C_{4B})} \cdot \frac{2C_{3A}}{(C_{2A} + 2C_{3A} + C_{4A})} \cdot \frac{C_{2B}}{(C_{2B} + 2C_{3B} + C_{4B})} \qquad (39)$$

and, in general,

$$P_{cA,i} =$$

$$\frac{C_{2B}}{(C_{2B} + 2C_{3B} + C_{4B})} \cdot \left(\frac{2C_{3B}}{(C_{2B} + 2C_{3B} + C_{4B})} \cdot \frac{2C_{3A}}{(C_{2A} + 2C_{3A} + C_{4A})} \right)^{i-1} \qquad (40)$$

With the preceding interpretation of the terms in eqn. 35, these equations and the corresponding equations for B—B units become

$$\frac{1}{k_c}\frac{dC_{1A}}{dt} = -2C_{1A}(2C_{1B} + C_{2B})$$

$$\frac{1}{k_c}\frac{dC_{2A}}{dt} = 2C_{1A}(2C_{1B} + C_{2B}) - C_{2A}(2C_{1B} + C_{2B})$$

$$- \text{Pab}\,.\,C_{2A}\cdot\frac{C_{2B}}{(C_{2B} + 2C_{3B} + C_{4B})}$$

$$\times \sum_{i=1}^{\infty}\left(\frac{2C_{3B}}{(C_{2B} + 2C_{3B} + C_{4B})}\cdot\frac{2C_{3A}}{(C_{2A} + 2C_{3A} + C_{4A})}\right)^{i-1}\cdot\frac{1}{i^{3/2}}$$

$$\frac{1}{k_c}\frac{dC_{3A}}{dt} = C_{2A}(2C_{1B} + C_{2B})$$

$$\frac{1}{k_c}\frac{dC_{4A}}{dt} = \text{Pab}\,.\,C_{2A}\cdot\frac{C_{2B}}{(C_{2B} + 2C_{3B} + C_{4B})}$$

$$\times \sum_{i=1}^{\infty}\left(\frac{2C_{3B}}{(C_{2B} + 2C_{3B} + C_{4B})}\cdot\frac{2C_{3A}}{(C_{2A} + 2C_{3A} + C_{4A})}\right)^{i-1}\cdot\frac{1}{i^{3/2}}$$

$$\frac{1}{k_c}\frac{dC_{1B}}{dt} = -2C_{1B}(2C_{1A} + C_{2A})$$

$$\frac{1}{k_c}\frac{dC_{2B}}{dt} = 2C_{1B}(2C_{1A} + C_{2A}) - C_{2B}(2C_{1A} + C_{2A})$$

$$- \text{Pab}\,.\,C_{2B}\cdot\frac{C_{2A}}{(C_{2A} + 2C_{3A} + C_{4A})}$$

$$\times \sum_{i=1}^{\infty}\left(\frac{2C_{3A}}{(C_{2A} + 2C_{3A} + C_{4A})}\cdot\frac{2C_{3B}}{(C_{2B} + 2C_{3B} + C_{4B})}\right)^{i-1}\cdot\frac{1}{i^{3/2}}$$

$$\frac{1}{k_c}\frac{dC_{3B}}{dt} = C_{2B}(2C_{1A} + C_{2A})$$

$$\frac{1}{k_c}\frac{dC_{4B}}{dt} = \text{Pab}\,.\,C_{2B}\cdot\frac{C_{2A}}{(C_{2A} + 2C_{3A} + C_{4A})}$$

$$\times \sum_{i=1}^{\infty}\left(\frac{2C_{3A}}{(C_{2A} + 2C_{3A} + C_{4A})}\cdot\frac{2C_{3B}}{(C_{2B} + 2C_{3B} + C_{4B})}\right)^{i-1}\cdot\frac{1}{i^{3/2}}$$

$$(41)$$

This set of simultaneous differential equations is effectively that used by Gordon and Temple,[46] with the derivatives of zero-order generating functions expressed as concentrations and those of first-order generating functions as ratios of concentrations. In general, for an $RA_{f_a} + RB_{f_b}$ random polymerisation $(f_a(f_a + 1)/2) + (f_b(f_b + 1)/2) + 2$ such equations will arise. This represents a large reduction in the number of equations used in the kinetics approach. Solution of the equations gives the concentrations of the various states as functions of $k_c t$, for given values of Pab, C_A and C_B, the last two quantities being defined by the reaction mixture. The initial conditions are $C_{1A} = C_A$, $C_{1B} = C_B$, and $C_{iA} = C_{iB} = 0$ for $i > 1$. For the $RA_2 + RB_2$ (linear) polymerisation the sums in eqn. 41 converge rapidly.

To interpret total ring-fraction data, the overall concentrations of groups which are unreacted (c_ω) and those which have reacted inter-molecularly (c_α) and intra-molecularly (c_σ) are required. For the A—A units these are given by the equations

$$c_{\omega a} = 2C_{1A} + C_{2A} \tag{42}$$

$$c_{\alpha a} = C_{2A} + 2C_{3A} + C_{4A} \tag{43}$$

$$c_{\sigma a} = C_{4A} \tag{44}$$

As the total concentration of A groups is $2C_A$, these concentrations convert to the corresponding fraction of unreacted groups (ω) and extents of reaction (α and σ) on division by this quantity, giving

$$\omega_a = (2P_{1A} + P_{2A})/2 \tag{45}$$

$$\alpha_a = (P_{2A} + 2P_{3A} + P_{4A})/2 \tag{46}$$

and

$$\sigma_a = P_{4A}/2 \tag{47}$$

Equations 45 to 47 satisfy the normalisation condition

$$\omega_a + \alpha_a + \sigma_a = 2\sum_i P_{iA}/2 = 1 \tag{48}$$

In addition,

$$\alpha_a + \sigma_a = p_a \tag{49}$$

the conventional, overall extent of reaction of A groups. Further, the concentration of molecules in the reaction mixture is the number initially less than the number lost through inter-molecular reaction, i.e.

$$C_A + C_B - 2C_A\alpha_a \quad \text{or} \quad C_A + C_B - 2C_B\alpha_b \tag{50}$$

and the concentration of rings is C_{4A} or C_{4B}. Thus,

$$N_r = \frac{C_{4A}}{C_A + C_B - 2C_A\alpha_a} = \frac{C_{4B}}{C_A + C_B - 2C_B\alpha_b} \tag{51}$$

and use of eqns. 42 to 47 shows that

$$N_r = \frac{\sigma_a}{\left(\dfrac{1}{2} + \dfrac{1}{2r} - \alpha_a\right)} = \frac{\sigma_b}{\left(\dfrac{r}{2} + \dfrac{1}{2} - \alpha_b\right)} \tag{52}$$

where r is the usual initial ratio of reactive groups, i.e. $r = 2C_A/2C_B$. For an equimolar reaction mixture, $\sigma_a = \sigma_b = \sigma$, etc., and

$$N_r = \sigma/(1 - \alpha) \tag{53}$$

as in eqn. 34. Thus, eqns. 41 are solved with $k_c t$ as independent variable to give σ and α, or σ_a and α_a, which are then combined to give N_r as a function of $p\, (= \alpha + \sigma)$ or p_a. Thus, N_r versus p curves may be constructed for given values of Pab and compared with the experimental curves to deduce the best values of Pab and b to describe the data.

6.2. Rate Theory

This was derived independently of cascade theory.[4,51] It also considers subsets of states of monomer units but adopts the traditional statistical approach of Flory[1] to random polymerisation, defining existence probabilities of molecular species rather than just those of structural units in terms of the functional groups. Given the existence probabilities of molecular species in terms of extents of reaction without ring formation, the rate theory evaluates the rates at which the species interconvert as functions of extents of reaction. Intra-molecular reaction is introduced as a perturbation to the resulting set of *independent* (rather than simultaneous) differential equations with the sums of existence probabilities still normalised to unity. The rate theory without intra-molecular reaction has been defined in general terms for linear random polymerisations[4] and has been extended to $RA_2 + RB_2$ polymerisations including intra-molecular reaction.[51] In particular, the limiting form of the theory, in which the distinction between p and p' is neglected and only the smallest rings are considered, has been applied to the prediction of total ring fraction data. Its further applications, to intra-molecular reaction in non-linear random polymerisations, gelation and network formation, are in progress.[63,81]

The essential features of the approach and its treatment of intra-molecular reaction can most easily be seen by considering the limiting form of the description of an RA_2 polymerisation. The subset of states of a monomer unit is first summarised in Table 3. The subset may be extended, always subject to the conditions $\Sigma P_i = 1$ and $\Sigma dP_i/dp = \Sigma P_i' = 0$. Also, general rules have been defined[4] for constructing the terms in the rate expressions *without* differentiation. Negative terms signify the disappearance and positive terms the appearance of a given species. The existence probabilities may be generated from the rates of change by the integration of individual terms. Thus, as the initial conditions are $P_1 = 1$, and $P_i = 0, i > 1$,

$$P_1 = 1 - \int_0^p 2(1-p)dp = 1 - P_{12} = 1 - p(2-p) \tag{54}$$

$$P_2 = \int_0^p 2(1-p)dp - \int_0^p 2pdp = P_{12} - P_{23} = p(2-p) - p^2 \tag{55}$$

and

$$P_3 = \int_0^p 2pdp = P_{23} = p^2 \tag{56}$$

where P_{12} is the total fraction of units which has been converted from state 1, and P_{23} the fraction which has been converted to state 3. Thus, the overall description of the polymerisation process goes beyond that afforded by the classical statistical description of random polymerisations, allowing not only existence probabilities (which are in fact weight or unit fractions[4]) to be described but also how they interconvert. The description is illustrated in Fig. 8 for the states and routes in Table 3 and eqns. 54 to 56.

TABLE 3

RATE THEORY:[4] SMALLEST SUBSET OF STATES OF AN <u>A—A</u> UNIT IN AN RA_2
POLYMERISATION TOGETHER WITH EXISTENCE PROBABILITIES AND THEIR RATES OF
CHANGE[a]

State	Existence probability	Rate of change
1 <u>A—A</u>	$P_1 = (1-p)^2$	$P_1' = -2(1-p)$ $\qquad - (1,2)$
2 <u>A—A</u>A—	$P_2 = 2p(1-p)$	$P_2' = 2(1-p) - 2p$ $\qquad + (1,2)\ -(2,3)$
3 —<u>A</u>A—<u>A</u>A—	$P_3 = p^2$	$P_3' = 2p$ $\qquad + (2,3)$

[a] The numbers in parentheses give the routes by which states interconvert, e.g. (i,j) specifies route i,j or the reaction of state i to form state j.

For the subset in Table 3, only state 1 can undergo intra-molecular reaction, when the subset becomes modified to that in Table 4. With respect to P'_1, c_a is the instantaneous external concentration of A groups, equal to $c_{ao}(1-p)$, and Pab is the internal concentration for forming the smallest ring. The factor $2c_a$ arises because of the two sites for inter-molecular reaction. In general, for any state which can undergo intra-molecular reaction, it is assumed that its *total* rate of disappearance with respect to p is unchanged. However, the number of individual routes is increased and their rates are redistributed according to the instantaneous probabilities of the various reactions involved (cf. eqn. 28).

The amounts interconverted by various routes are most easily obtained by numerical rather than analytical integration of the respective rates. For example, with larger subsets or $RA_2 + RB_2$ polymerisations, lengthy expressions have to be derived if analytical integration is used. Thus, in Table 4, the existence probabilities have merely been expressed in terms of amounts which have appeared and disappeared via the various routes. However, unlike the previous theories, it is possible to derive analytical expressions for the existence probabilities. For example, from Table 4, the unit fraction of rings

$$R_4 = \int_0^p 2(1-p) \cdot \frac{\text{Pab}}{2c_a + \text{Pab}} \, dp$$

giving

$$R_4 = \lambda p + \frac{\lambda^2}{2} \ln \left(\frac{2(1-p) + \lambda}{2 + \lambda} \right) \tag{57}$$

where

$$\lambda = \text{Pab}/c_{a0} \ (= c_{\text{int},1}/c_{0,\text{ext}}) \tag{58}$$

is a ring-forming parameter defined by the initial concentration of reactive groups $(c_{0,\text{ext}})$ and, through Pab, by the monomer or prepolymer chain structure and chain length.

Figure 9 illustrates the descriptions of the polymerisation according to Tables 3 and 4. Solid curves are as in Fig. 8 and the dashed curves are those resulting from the effects of intra-molecular reaction with $\lambda = 0.4$. Notice that the amount of state 1 which has reacted to form state 2 has been reduced by intra-molecular reaction, whereas the amount which has disappeared from state 2 is initially unchanged. Thus, as p approaches 1, P_2 becomes negative. This is because the perturbation to the rates to include intra-molecular reaction has been written in terms of the overall extent of

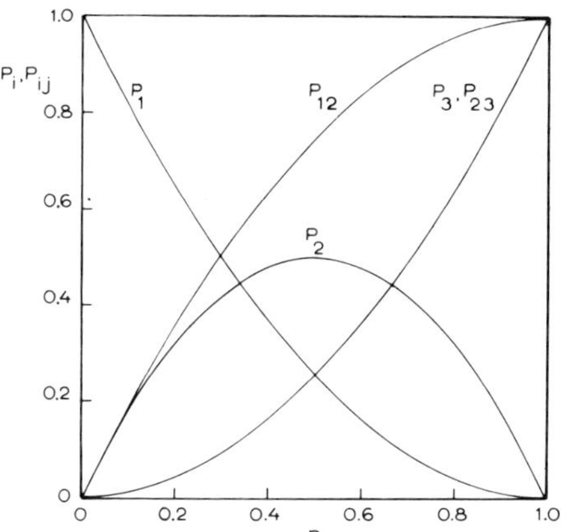

FIG. 8.   Rate theory.[4] Existence probabilities (P_i) and amounts which have appeared and disappeared via individual routes (P_{ij}) for no intra-molecular reaction and the smallest subset of states in an RA_2 polymerisation. (See Table 3 and eqns. 54 to 56.)

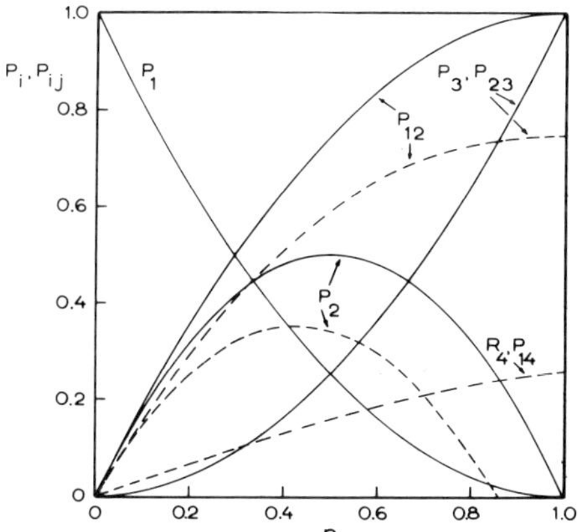

FIG. 9.   Rate theory.[4,51] As Fig. 8, but including intra-molecular reaction with ring-forming parameter $\lambda = 0.4$. (See Tables 3 and 4 and eqns. 54 to 58.) ———, Curves in the absence of intra-molecular reaction $(P_2, P_{12},$ and P_3 or $P_{23})$ or unaffected by intra-molecular reaction (P_1); ———, curves affected by intra-molecular reaction $(P_2, P_{12}, P_3$ or P_{23} for $p > 0.86$, and R_4 or $P_{14})$.

TABLE 4

RATE THEORY:[4,51] PERTURBATION OF SUBSET OF STATES IN TABLE 3 TO INCLUDE INTRA-MOLECULAR REACTION

	State	Existence probability	Rate of change
1	A—A	$P_1 = 1 - P_{12} - P_{14}$	$P'_1 = -P'_{12} - P'_{14}$ $= -2(1-p) \cdot \dfrac{2c_a}{2c_a + \mathrm{Pab}} - 2(1-p) \cdot \dfrac{\mathrm{Pab}}{2c_a + \mathrm{Pab}}$ $-(1,2) \qquad -(1,4)$
2	A—AA—	$P_2 = P_{12} - P_{23}$	$P'_2 = P'_{12} - P'_{23}$ $= 2(1-p) \cdot \dfrac{2c_a}{2c_a + \mathrm{Pab}} - 2p$ $+(1,2) \qquad -(2,3)$
3	—AA—A A—	$P_3 = P_{23}$	$P'_3 = P'_{23} = 2p$ $+(2,3)$
4	A—A	$R_4 = P_{14}$	$R'_4 = P'_{14} = 2(1-p) \cdot \dfrac{\mathrm{Pab}}{2c_a + \mathrm{Pab}}$ $+(1,4)$

reaction rather than that in the chain fraction (p'). The procedure for evaluating existence probabilities with p' as variable has been described by Stanford *et al.*[51] In this case, the rate equations can only be integrated numerically as the relationship between p and p' has to be re-evaluated over small intervals in p'. However, a satisfactory interpretation of total ring fraction data in the $RA_2 + RB_2$ case has been possible using the limiting form of the theory (smallest subsets and $p = p'$).

Negative existence probabilities can in fact always be dealt with approximately within the limiting form of the theory.[51] They signify that more has disappeared from a state than has appeared. In the context of Table 4, this occurs when $P_{12} < P_{23}$. Thus, the point in the reaction when $P_{12} = P_{23}$ is noted ($p \cong 0{\cdot}86$ for $\lambda = 0{\cdot}4$), and beyond this point P_{23} is put equal to P_{12} so that state 2 remains exhausted ($P_2 = 0$). In other words, its rate of disappearance become equal to its rate of appearance, and, as it disappears to state 3, P'_3 becomes equal to P'_{12}. However, the treatment of negative existence probabilities is not needed for considerations of total ring fractions in linear random polymerisations, provided the smallest subsets are used. For example, in the present case, R_4 is unaffected.

The conversion of unit fraction (R_4) to number fraction of rings is easily achieved. The number of molecules present is the number expected without intra-molecular reaction plus the number of rings $= N_a(1-p) + N_a R_4$, where N_a is the number of RA_2 units in the polymerisation. Thus,

$$N_r = \frac{R_4}{(1-p) + R_4} \qquad (59)$$

allowing evaluation of N_r as a function of p for a given value of λ.

For $RA_2 + RB_2$ polymerisations, subsets for A—A and B—B units are defined and their existence probabilities are averaged according to the respective mole fractions of units,[4,51] to give unit fractions of molecular species as distinct from states. The smallest subsets contain six states without rings, the longest for the A—A subset being —AB—BA—A B— BA—, with the ring state, state 7, being $\overline{A—A\ B—B}$. For $r = 1$, for example, the rate expression for the unit fraction of rings, R_7, can be evaluated to give

$$R_7 = \lambda p(\lambda + p) + \frac{\lambda^2}{2}(2 + \lambda)\ln\left(\frac{2(1-p) + \lambda}{2 + \lambda}\right) \qquad (60)$$

and from reasoning similar to that leading to eqn. 59

$$N_r = \frac{R_7}{2(1-p) + R_7} \qquad (61)$$

6.3. Comparison of Analyses of Total Ring Fraction Data

The four analyses all essentially give calculated N_r versus p curves for given initial concentrations of reactants and chosen values of Pab, or b. The curves can then be compared with experimental ones. For example, Fig. 10 shows the experimental points and calculated curve for $b = 0.318$ nm using the kinetics analysis,[45] whilst Fig. 11 shows the same data, and calculated curves from the limiting form of the rate theory[51] with the stated values of λ, and with the curve at $\lambda = 0.6$ corresponding to $b = 0.33$ nm. In general, the kinetics analysis using the number of simultaneous equations previously stated was able to produce very good agreement between experimental points and calculated curves. This was also true of the rate theory, where, with reference to Fig. 11, closer overall agreement could be achieved by increasing slightly the value of λ. However, agreement at small p was sought as in this region the limiting form of the theory should more accurately apply.

The analysis according to cascade theory was found to agree with the kinetics analysis for the same values of b in the cases tried.[46] However, small but significant deviations between the two occurred for larger numbers of rings, indicating that the spanning-tree approximation has a tendency to

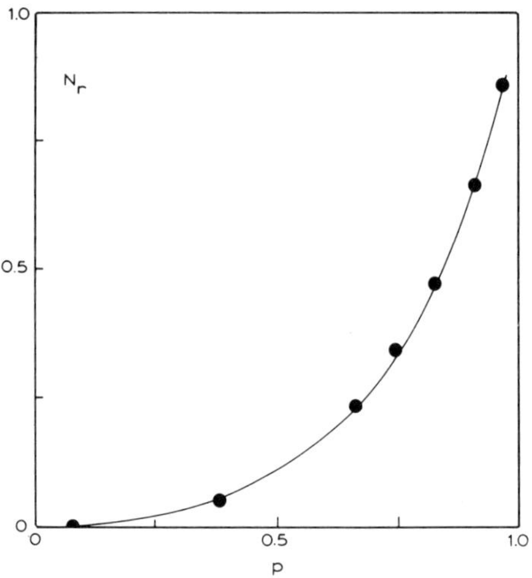

FIG. 10. Number fraction of rings (N_r) versus extent of reaction (p). Fitting of data from experiment 4 of Table 1 according to kinetics analysis.[45] ●, Experimental points (see Fig. 6); ———, calculated curve using effective bond length $b = 0.318$ nm. (Reproduced from Gordon, M. and Temple, W. B., *Makromol. Chem.*, **152** (1972), p. 277, by permission of Hüthig and Wepf Verlag, Basel ©.)

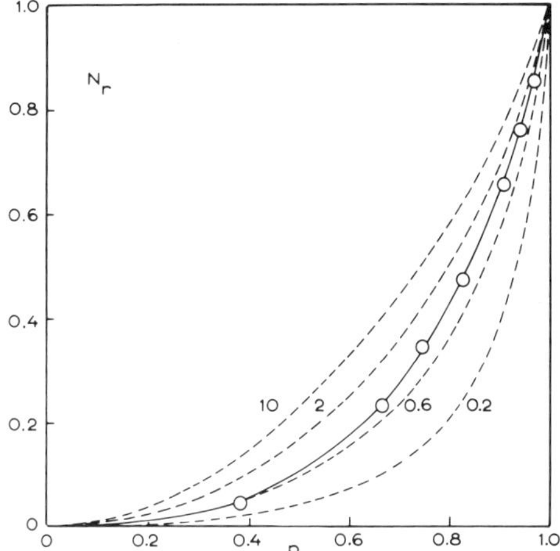

FIG. 11.   Plots of number fraction of rings (N_r) versus extent of reaction (p) for equimolar reaction mixtures ($r = 1$), according to limiting form of rate theory for $RA_2 + RB_2$ polymerisations.[51] $\bigcirc$, Experimental points from experiment 4 of Table 1 and Fig. 6; – – –, calculated curves at stated values of ring-forming parameter λ. (Reproduced from Stanford, J. L. *et al.*, *J. Chem. Soc., Faraday Trans. I*, **71** (1975), p. 1308, by permission of Royal Society of Chemistry ©.)

underestimate ring formation. The predictions of the two analyses are compared in Fig. 12, where the points correspond to *calculated* values of N_r from cascade theory and the curves to *calculated* values from the kinetics analysis according to the initial concentrations of experiments 2, 4 and 5 of Table 1.

Finally, an example of the curve-fitting according to the more approximate Jacobson–Stockmayer–Kilb (JSK) analysis[44] of eqns. 30 and 32 is shown in Fig. 13 for the data from experiment 2 of Table 1 and Fig. 6. The values of λ used are given with the calculated curves and the data are presented as $R/(C_a + C_b)$ versus p (see eqn. 30). Generally, only approximate agreement with the experimental data was found.

A summary of the published results of the four analyses is given in Table 5 according to the values of b which gave best fits to the experimental data. With reference to the experiments at $r = 1$, the rate theory gives an essentially constant value for b for experiments 2, 3 and 4 and a smaller value for experiment 6 which used HDI/PEG400. This is in accord with the larger proportion of ethylene oxide units in the HDI/PEG400 chain as compared with the HDI/PEG200 chain of experiments 2, 3 and 4. A

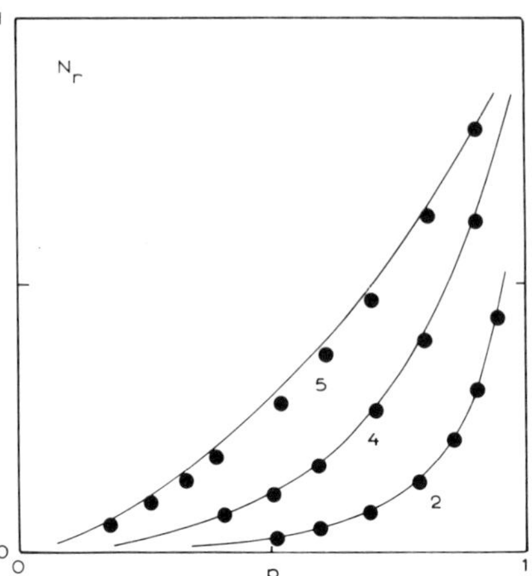

FIG. 12. Number fraction of rings (N_r) versus extent of reaction (p). Comparison[46] of prediction of cascade theory ($\bullet$) and kinetics analysis (———). Calculations for the initial concentrations of reactants given in Table 1 for experiments 2, 4 and 5, as indicated with the curves, using effective bond length $b = 0\cdot368$ nm, $0\cdot318$ nm and $0\cdot318$ nm, respectively. (Reproduced from Gordon, M. and Temple, W. B., *Makromol. Chem.*, **160** (1972), p. 263, by permission of Hüthig and Wepf Verlag, Basel ©.)

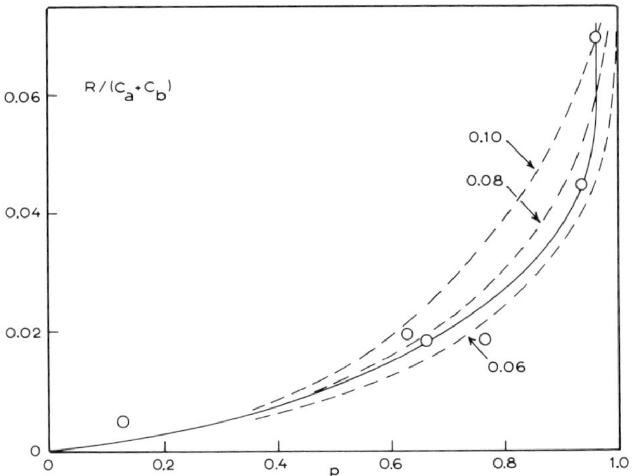

FIG. 13. Number of rings per monomer unit, $R/(C_a + C_b)$, versus extent of reaction (p) for experiment 2 of Table 1 and Fig. 6.[44] ———, Curve through experimental points; – – –, Jacobson–Stockmayer–Kilb (JSK) analysis at the values of ring-forming parameter λ indicated. (Reproduced from Stepto, R. F. T. and Waywell, D. R., *Makromol. Chem.*, **152** (1972), p. 263, by permission of Hüthig and Wepf Verlag, Basel ©.)

TABLE 5

VALUES OF b DERIVED FROM ANALYSES OF TOTAL RING FRACTION DATA FROM LINEAR POLYURETHANE-FORMING REACTIONS

Experiment	Percentage of monomers (w/w)	r^a	b (nm)			
			JSK analysis	Rate theory analysis	Kinetics analysis	Cascade theory analysis
1	100	1	—	—	0·380	—
2	20	1	0·39	0·33	0·368	0·368
3	10	1	0·39	0·34	0·348	—
4	5	1	0·39	0·33	0·318	0·318
5	1	1	—	—	<0·318	<0·318
6[b]	5	1	0·33	0·29	0·318	—
7	5	1/0·64	—	0·30	0·318	—
8	5	1/0·33	—	0·28	0·292	—
9	5	0·60/1	—	0·31	0·318	—
10	5	0·33/1	—	—	—	—

[a] $r = [NCO]_0/[OH]_0$.
[b] HDI/PEG400, other experiments HDI/PEG200.
(Reproduced in part from Stanford, J. L., Stepto, R. F. T. and Waywell, D. R., *J. Chem. Soc., Faraday Trans. I*, **71** (1975), p. 1308, by permission of Royal Society of Chemistry ©.)

corresponding variation of b with chain structure has been found from analyses of gelation data—see section 9. Less significantly, the more approximate JSK analysis gave a similar variation in b.

The values of b from the rate theory are representative of the chains forming the smallest rings. They may be compared with the values of b for infinite polymethylene and polyoxyethylene chains,[66] namely 0·4 nm and 0·29 nm, respectively. The fact that they tend to lie towards the lower end of the range defined by these two values may be attributed to the fact that b increases with chain length. Also, the form of the theory used accounted only for the smallest rings and any undercounting would require larger values of Pab (or smaller values of b) in compensation. However, as mentioned with respect to Fig. 11, the values of b were selected to give agreement at small p, where the distribution of ring sizes may be expected to be sensibly constant and hopefully to be dominated by the smallest rings. Thus, the values of b deduced would appear to have some significance in terms of chain flexibility, subject of course to the additional assumption of Gaussian statistics at $\mathbf{r} \cong \mathbf{0}$ (see eqn. 15 and ensuing discussion).

In contrast, the values of b from the kinetics (and cascade) analysis do not reflect any difference in b for HDI/PEG200 and HDI/PEG400. They also decrease with dilution. However they do refer to the whole range of values of p, and for different dilutions the different values of b obtained could reflect differences in the distributions of ring sizes. The value of b for each experiment will refer to some average ring size, $\langle i \rangle$, and the results indicate that $\langle i \rangle$, and, hence, b decrease with dilution. In other words, a larger proportion of smaller rings is formed as dilution increases.[51] An alternative explanation for the decrease in b in terms of chain contraction on dilution has been suggested.[45]

The values of b from the analyses of experiments at $r \neq 1$ are less easy to understand. However the variations in b found for the experiments, all at 5% monomers, could again be due in some part to changes in ring distributions (this time with r). In addition, the occurrence of side reactions in the presence of a large excess of isocyanate will have an effect (see Fig. 7 and ensuing discussion).

The preceding detailed discussion of the values of b in Table 5 should not mask the fact that from the kinetics analysis, and the cascade and rate theory analyses, it is possible to fit experimental total-ring-fraction curves. In contrast with analyses of ring-chain equilibria (section 3), detailed calculations of Pab using R–I–S models have not been carried out to provide independent estimates of b for chains of a given length. For ring-chain equilibria much more detailed data are available and such calculations are justified. For the irreversible case, such detailed data, the concentrations of *individual* ring species for several systems, are required before a more critical examination of the theories can be undertaken.

Of the approaches used, the cascade theory and the rate theory are of more general use in descriptions of irreversible random polymerisations. They require fewer equations to be solved and can be more easily applied to non-linear polymerisations. They both use subsets of states of the monomer units. However, cascade theory employs 'pseudo-time' as an independent variable and simultaneous differential equations, whereas rate theory uses extents of reaction and independent differential equations. Such independent equations lead in principle to analytical expressions for ring fractions (e.g. eqns. 57, 59, 60 and 61), but in practice numerical solutions may be preferred due to the complicated expressions which need to be integrated. The rate theory also allows different levels of approximation to be clearly defined, e.g. size of subset and use of p or p', but the effects of the spanning-tree approximation in cascade theory can be difficult to quantify.

7. PRE-GEL INTRA-MOLECULAR REACTION IN IRREVERSIBLE NON-LINEAR RANDOM POLYMERISATION

Stanford and Stepto[60] have extended the experimental work[44] on linear polyurethane-forming systems and studied intra-molecular reaction in $RA_2 + RB_3$ polymerisations. The same experimental techniques were used to determine N_r and reactions were again carried out at $70\,°C$ in bulk and at various dilutions in benzene using different initial ratios of reactive groups. The reactants were HDI and a polyoxypropylene (POP) triol (LG56) with the structural formula

$$CH_2\!\!-\!\![OCH_2CH(CH_3)]_{\bar{n}}OH$$
$$|$$
$$CH\ -\!\![OCH_2CH(CH_3)]_{\bar{n}}OH$$
$$|$$
$$CH_2\!\!-\!\![OCH_2CH(CH_3)]_{\bar{n}}OH$$

In the formula, $\bar{n}$, the number-average d.p. of each arm, was approximately equal to 17 for the two batches of triol used, giving $v = 115$, as compared with 25·2 for the linear HDI/PEG200 system; v is defined as the average chain length of two triol arms and an HDI residue. Triol samples were characterised by end-group analysis and M_n determinations showing (number-average) hydroxyl functionalities of 2·99 and 2·95. Again, independent kinetics investigations[8,90,91] had shown that the hydroxyl groups had equal reactivities.

The quantity N_r was evaluated from measured values of M_n according to eqn. 29, with $f_a = 2$, $f_b = 3$ and $r = [NCO]_0/[OH]_0$. In a non-linear polymerisation, N_r is the average number of *ring structures* per molecule, rather than number fraction of rings, as molecules containing intra-molecularly reacted groups can undergo further reaction, as illustrated in Fig. 14.

Figure 15 shows N_r versus p for reactions at $r = 1$. The curves stop before the gel-points of the respective reactions. As the initial concentration of reactive groups increases, N_r decreases. However, even in bulk (experiment 1.1) about one molecule in three contains a ring structure at the gel-point ($p_{gel} = 0·765$, and $N_r \cong 0·3$). This value is in marked contrast to the bulk linear system studied by Stepto and Waywell,[44] for which $N_r \cong 0$ although the value of v was much smaller (see Fig. 6). The underlying reason is that for a non-linear random polymerisation the number of reactive groups per molecule increases as the polymerisation proceeds. There are many more pairs of groups which can undergo intra-molecular

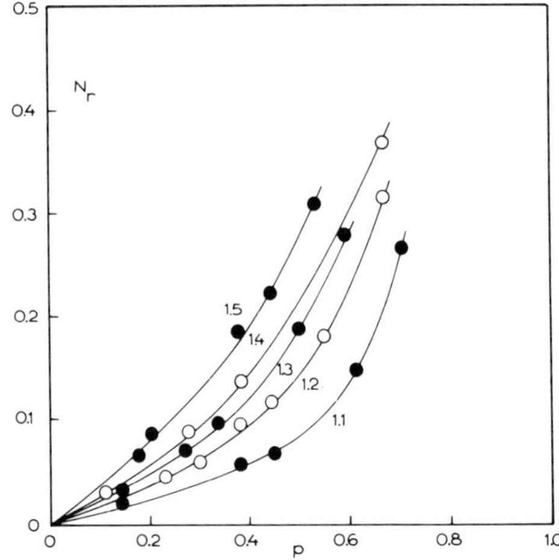

FIG. 14. Examples of ring structures in an $RA_2 + RB_3$ polymerisation.[60] (a) Pentamer with functional groups completely reacted. (b) Part of a continuing molecule.(Reproduced from Stanford, J. L. and Stepto, R. F. T., *Brit. Polym. J.*, **9** (1977), p. 124, by permission of the authors).

reaction and the effect of these is to outweigh the reduction in probability that a given pair of groups will so react. The reduction in probability arises from the value $v = 115$ as compared with 25·2 and probably a larger value of b for the HDI/POP chain in comparison with the HDI/PEG200 chain.

Figure 16 further compares data from linear and non-linear polymerisations and also illustrates the effects of reactant ratio. All the data refer to

FIG. 15. Number of ring structures per molecule (N_r) versus extent of reaction (p) for equimolar reaction mixtures ($r = 1$).[60] Non-linear $RA_2 + RB_3$ polymerisations, HDI/POP triol (LG56)/70°C in bulk and at various dilutions in benzene. ●, experiment 1.1, bulk; ○, experiment 1.2, 70% monomers; ●, experiment 1.3, 50% monomers; ○, experiment 1.4, 40% monomers; ●, experiment 1.5, 30% monomers. LG56: oxypropylated glycerol, as illustrated in text. (Reproduced from Stanford, J. L. and Stepto, R. F. T., *Brit. Polym. J.*, **9** (1977), p. 124, by permission of the authors.)

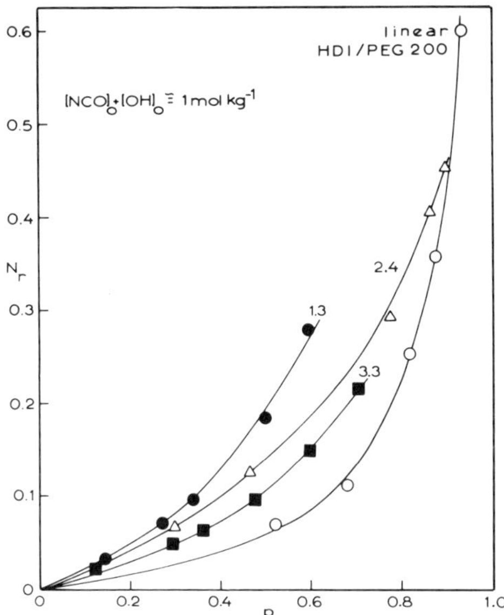

Fig. 16. Number of ring structures per molecule (N_r) versus extent of reaction (p), for non-linear and linear random polymerisations at the same initial concentration. ●, Experiment 1.3, $r = 1$, 50% monomers, $[NCO]_0 + [OH]_0 = 0.8975\,mol\,kg^{-1}$. △, Experiment 2.4, $r = 1.5$, 40% monomers, $[NCO]_0 + [OH]_0 = 0.8806\,mol\,kg^{-1}$. ■, Experiment 3.3, $r = 1/1.5$, 58% monomers, $[NCO]_0 + [OH]_0 = 0.9111\,mol\,kg^{-1}$. ○, HDI/PEG200, experiment 3 of Table 1, $r = 1$, 10% monomers, $[NCO]_0 + [OH]_0 = 1.0616\,mol\,kg^{-1}$.[44] (Reproduced from Stanford, J. L. and Stepto, R. F. T., *Brit. Polym. J.*, **9** (1977), p. 124, by permission of the authors.)

reaction systems in which $[NCO]_0 + [OH]_0 \cong 1\,mol\,kg^{-1}$. Comparison of curves 1.3 (at $r = 1$) and that for the linear polymerisation clearly indicates the increased opportunities for intra-molecular reaction in non-linear polymerisations. Also, curve 1.3 is for 50% monomers whilst that for the linear system is for 10% monomers, emphasising that for considerations of intra-molecular reaction it is the initial concentrations of reactive groups and not the percentage by weight of solvent which is of prime importance.

For the non-linear systems most ring structures are again formed for $r = 1$ (curve 1.3). The curves at reciprocally related values of r (2.4 and 3.3) are not exactly coincident. This may be due to the facts that the values of r were not exactly reciprocally related (1.497 and 1/1.503) and that expt 2.4 was at a slightly lower initial concentration of reactive groups. However, it may be that in non-linear polymerisations equal opportunities for intra-molecular reaction do not arise for reciprocally related values of r. If so, curves 2.4 and 3.3 indicate that an excess of bifunctional reagent gives more intra-molecular reaction. There is some support for this conclusion from gelation studies.[61] However, as the results for linear systems also indicate, further investigations of the effects of ratio are needed.

The quantitative interpretation of the data of Stanford and Stepto[60] according to cascade theory and rate theory is at present in progress.[63,81] Temple[47] devised a kinetics analysis of $RA_2 + RB_3$ polymerisations including intra-molecular reaction. However, the complexity of molecular species occurring means that such a description is limited to low extents of reaction ($p \cong 0.3$). The values of σ predicted were compared with those from cascade theory and indicated that the latter underestimates intra-molecular reaction.

8. GELATION IN NON-LINEAR RANDOM POLYMERISATIONS

Given the equal reactivity of like functional groups and the absence of intra-molecular reaction, the gel-point in an $RA_{f_a} + RB_{f_b}$ polymerisation occurs when

$$p_a p_b (f_a - 1)(f_b - 1) = 1 \tag{62}$$

This equation stems from the more general relationship for the case when several reactants, some bearing A groups and some bearing B groups, are involved,[7] namely,

$$p_a p_b (f_{aw} - 1)(f_{bw} - 1) = 1 \tag{63}$$

Here, f_{aw} and f_{bw} are the weight-average functionalities of the reactants bearing A groups and B groups, respectively. Equations 62 and 63 state that there is a point in the polymerisation (the gel-point) where there is unit probability that the 'repeating structure' of the polymer formed can continue *ad infinitum*. From this point onwards, molecules limited in size only by the macroscopic amount of material in the reaction mixture make their appearance, so that sol and gel fractions coexist, consisting of finite, and macroscopic or network species, respectively.[1c]

Equation 63 can be simply derived,[23] and in a manner which illustrates the usefulness of the concepts of repeating structure and f_w. Consider a reaction mixture consisting of N_{a1}, N_{a2}, ... moles of reactants bearing f_{a1}, f_{a2} ... A groups reacting with N_{b1}, N_{b2} ... moles of reactants bearing f_{b1}, f_{b2} ... B groups. Choose an A group at random from the mixture, as illustrated on the left of Fig. 17. The probability that it belongs to the monomer of functionality f_{ai} is given by the expression

$$N_{ai} \cdot f_{ai}/\Sigma N_{ai}f_{ai} = f_a(i) \tag{64}$$

the fraction of A groups which are on units of that functionality. The probability that the unit to which the chosen A group is attached is connected to a unit bearing B groups is $(f_{ai} - 1)p_a$. There are $(f_{ai} - 1)$ possible paths to a unit bearing B groups and each has a probability p_a of continuing. Assume that, as illustrated, one has continued, then there is a

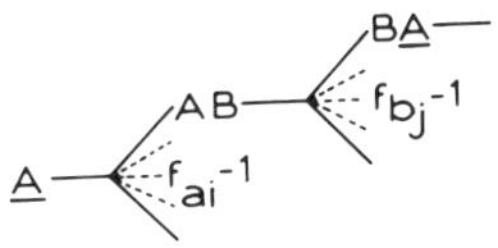

FIG. 17.   Repeating structure $\underline{A}$ to $\underline{A}$, for $RA_{f_{ai}}$ and $RB_{f_{bj}}$ units in a general non-linear random polymerisation.

probability $f_b(j)$ (corresponding to $f_a(i)$) that it joins to a unit bearing j B groups. This unit can in turn lead to an A unit with probability $(f_{bj} - 1)p_b$, thus spanning the repeating structure defined by $RA_{f_{ai}}$ and $RB_{f_{bj}}$ units. Hence the probability that this structure is spanned, is

$$f_a(i)(f_{ai} - 1)p_a \cdot f_b(j)(f_{bj} - 1)p_b \tag{65}$$

The average repeating structure of the whole reaction mixture includes all species i and j and the corresponding probability of continuation is

$$\sum_{i,j} f_a(i)(f_{ai} - 1)p_a \cdot f_b(j)(f_{bj} - 1)p_b \tag{66}$$

This expression factorises into a product of sums over i and j, giving for gelation

$$p_a p_b \left(\sum_i f_{ai}f_a(i) - \sum_i f_a(i) \right) \cdot \left(\sum_j f_{bj}f_b(j) - \sum_j f_b(j) \right) = 1 \tag{67}$$

126 R. F. T. STEPTO

However,

$$\sum_{i} f_{ai} f_a(i) = \sum_{i} N_{ai} \cdot f_{ai}^2 \bigg/ \sum_{i} N_{ai} f_{ai} = f_{aw} \tag{68}$$

$$\sum_{j} f_{bj} f_b(j) = \sum_{j} N_{bj} f_{bj}^2 \bigg/ \sum_{j} N_{bj} f_{bj} = f_{bw} \tag{69}$$

and

$$\sum_{i} f_a(i) = \sum_{j} f_b(j) = 1 \tag{70}$$

giving immediately eqn. 63.

A simple application of eqn. 63 is for an $RA_2 + RA_f + RB_2$ polymerisation. Flory[1c,5] has derived an expression for the gel-point giving

$$p_a p_b ((2(1-\rho) + f_\rho) - 1) = 1 \tag{71}$$

where ρ is the fraction of A groups on RA_f units. In fact this equation follows directly from eqn. 63 as $f_{bw} = 2$ and $f_{aw} = 2(1-\rho) + f\rho$.

Although, given equal reactivity and no intra-molecular reaction, the Flory–Stockmayer treatment of gelation is really established beyond doubt, references unfortunately still seem to occur in the literature (e.g. refs. 76, 101, 102) to the earlier treatment of Carothers,[103] according to which gelation occurs when all the monomer units could be joined together. For example, in an RA_f polymerisation, this approach leads incorrectly to the gel-point occurring when $p = 2/f$. In addition, independently derived theories using this concept have more recently been proposed.[104,105] The reader is referred to the literature[106–109] for the ensuing discussions.

The data of Stanford and Stepto[60] on total ring fractions clearly shows that pre-gel intra-molecular reaction cannot normally be neglected in gelling systems. The effect of such reaction in delaying the gel-point (increasing the value of $p_a p_b$ at gelation) can be seen from the work of Hopkins, Peters and Stepto[8] using similar di-isocyanate/triol systems. Reactions were carried out at 80 °C in bulk and in nitrobenzene solution. HDI and decamethylene di-isocyanate (DDI) and three POP triols of different molar masses were used. Reactions were carried out with $r \simeq 1$. Figure 18 shows the results obtained as a function of initial dilution of the reaction mixtures. The critical value of $p_{OH} \cdot p_{NCO}$ at the gel-point is denoted by α_c. The curves drawn are consistent with no intra-molecular reaction in

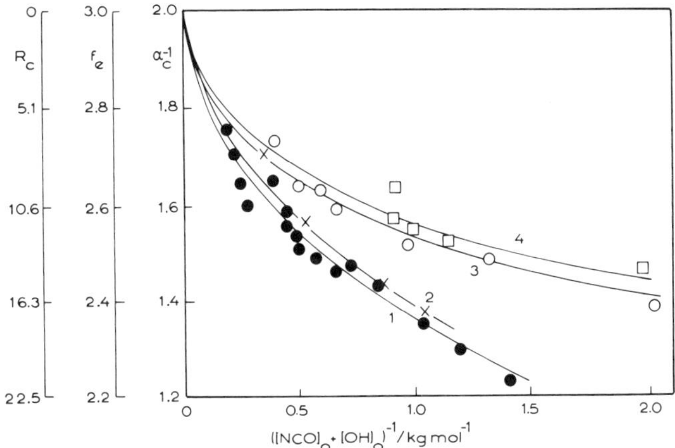

FIG. 18.   Reciprocal product of extents of reaction at gelation (α_c^{-1}) versus initial dilution of reaction groups $(([NCO]_0 + [OH]_0)^{-1})$. The alternative ordinates are discussed in the text. Di-isocyanate/triol systems:[8] 1, HDI/LHT240, $v = 36$; 2, DDI/LHT240, $v = 40$; 3, HDI/LHT112, $v = 66$; 4, HDI/LG56, $v = 112$; LHT240/ LHT112 are oxypropylated 1,2,6-hexane triols. (Reproduced from Hopkins, W. *et al.*, *Polymer*, **15** (1974), p. 315, by permission of IPC Business Press Ltd ©.)

the limit of zero dilution of reactive groups $((c_{ext})^{-1} = 0)$, triol functionalities of 3 and like groups having equal reactivities. Kinetics studies,[8,91,110] other methods of analysing the data[59,110] and independent analyses of two of the triols[60,110] indicate that their chemical functionalities are between 2·95 and 3. The kinetics studies have also shown that like groups had equal reactivities.

According to eqn. 62 gelation is expected to occur when $\alpha_c^{-1} = f_b - 1 = 2$. The results indicate that the measured value of α_c always exceeds the value of $\frac{1}{2}$ and increases with initial dilution of the reaction mixture. Noticeable deviations from this value are also apparent for the bulk systems, corresponding to the points shown at the lowest dilutions for the four reaction systems studied. In addition, at a given initial dilution, the deviation from the Flory–Stockmayer gel-point increases as the value of v characteristic of the reactants decreases. Thus, the behaviour displayed in Fig. 18 follows that expected from the considerations of intra-molecular reaction in the preceding sections. For a given system, the change of α_c^{-1} with dilution is consistent with the increase in pre-gel intra-molecular reaction with dilution shown in Fig. 15, whilst the variation in α_c^{-1} with v at a given dilution reflects the increased probabilities of intra-molecular reaction for smaller values of v or larger values of Pab.

The second ordinate in Fig. 18, f_e, gives the effective functionality which would be deduced assuming eqn. 62 was obeyed. The dangers of using such a functionality to infer the chemical functionality of the triol are apparent from the preceding discussion. Fogiel[41] and Guise and Zoch[79,80] have discussed the interpretation of effective functionalities derived from gel-point data and have illustrated the effects of intra-molecular reaction on such functionalities in a manner similar to that in Fig. 18.

The third ordinate in Fig. 18, R_c, is the percentage of intra-molecular reaction at gelation, defined[8] by the equation

$$R_c = 100(1 - ((f-1)\alpha_c)^{-\frac{1}{2}}) \tag{72}$$

for an $RA_2 + RB_f$ polymerisation at $r = 1$. It can be seen that α_c^{-1}, f_e, and R define alternative scales of the total amount of pre-gel intra-molecular reaction.

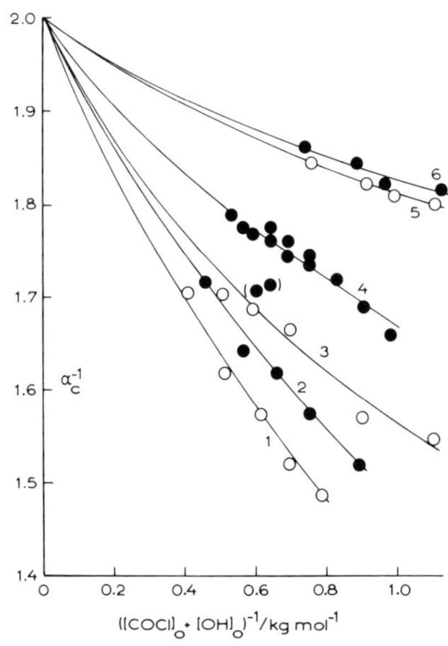

FIG. 19. Reciprocal products of extents of reaction at gelation (α_c^{-1}) versus initial dilution of reactive groups $(([COCl]_0 + [OH]_0)^{-1})$. Diacid chloride/triol systems:[62] 1, adipoyl chloride (AC)/LHT240, $v = 37$; 2, sebacoyl chloride (SC)/LHT240, $v = 41$; 3, AC/LHT112, $v = 66$; 4, SC/LHT112, $v = 70$; 5, AC/LG56, $v = 132$; 6, SC/LG56, $v = 136$. (Reproduced from Smith, R. S. and Stepto, R. F. T., *Makromol. Chem.*, **175** (1974), p. 2365, by permission of Hüthig and Wepf Verlag, Basel ©.)

Other studies of the effects of intra-molecular reaction on the gel-point avoiding the complication of like groups having unequal reactivities have been reported by Peters and Stepto[61] and Smith and Stepto.[62] The former work demonstrates the effects of initial dilution and ratio of reactants on the gel-point, again using an HDI/POP triol reaction system. As discussed with reference to Fig. 16, the results indicate that excess bifunctional component causes more intra-molecular reaction. The work of Smith and Stepto represents a detailed study of gel-points in related systems. Reactions between two diacid chlorides and three POP triols were used, giving six reaction systems with different values of v. Reactions at $r = 1$ were carried out at 60 °C with diglyme as solvent. The results are presented in Fig. 19, confirming the increase in intra-molecular reaction with increase in initial dilution and decrease in v.

9.   INTERPRETATIONS OF GELATION DATA

Several theoretical treatments concerning gelation in random poly-merisations which include intra-molecular reaction have been pub-lished.[9,19–22,27,33,36–38,41,48–50,55,56,58,59,81] Those developed from the cascade[19–22,33,49,50,55,56,58] and rate theories[81] seek to account for the development of intra-molecular reaction as a polymerisation proceeds and require numerical solutions to differential equations (cf. sections 6.1 and 6.2). Cascade theory has been applied to gelation data[19,20,50] and has been used to account for induced unequal reactivity along with intra-molecular reaction.[19,20] In addition, rate theory has recently been used in the interpretation of gelation data.[81] With both theories, values of the ring-forming parameter Pab can be derived which are consistent with experi-mental gel-points. However, it is not yet clear if they provide consistent interpretations of pre-gel intra-molecular reaction (N_r versus p) and α_c for the same system. The previously discussed work of Stanford and Stepto[60] contains α_c as well as N_r versus p data and is currently being used to examine this point.[63]

Most of the remaining theories[36,38,41,48,59] may be termed approximate theories of gelation.[59] They concentrate only on the gelation condition and, by making certain simplifying assumptions about the molecular structures existing during a polymerisation, result in analytical expressions which are easily used to interpret experimental data. Present considera-tions will be limited to those theories treating $RA_2 + RB_f$ polymerisations and to interpretations of the gelation data discussed in the preceding section.

130   R. F. T. STEPTO

Frisch[36] and Kilb[38] originally derived gel-point conditions which accounted for small amounts of intra-molecular reaction. Stepto[48] re-examined the expressions showing how they were related and provided an alternative derivation of Kilb's expression which avoided using that author's approximate modification of Jacobson–Stockmayer ring-chain equilibrium theory (see JSK analysis, section 6). The gel-point conditions of Frisch and Kilb are

$$\alpha_c(f-1)(1-(1-\alpha_c)\lambda_F) = 1 \qquad \text{(Frisch)} \qquad (73)$$

and

$$\alpha_c(f-1)(1-\lambda_K) = 1 \qquad \text{(Kilb)} \qquad (74)$$

λ_F and λ_K are ring-forming parameters to be determined from the experimental values of α_c, and, with the assumption of small amounts of intra-molecular reaction, interpreted[48] according to the equation

$$\lambda = \frac{c_{int}}{c_{ext}} \qquad (75)$$

where c_{int} is the *total* concentration of groups which can react intra-molecularly with a group about to react (see Fig. 1) and c_{ext} is the *instantaneous* concentration of groups which can react inter-molecularly with the same group; c_{ext} varies as the reaction proceeds. However, using approximate theories of gelation (which concentrate solely on the gel-point) there is no way of accounting for this variation. In other words, λ is taken as a constant and two extreme values for c_{ext} may be proposed,[48] namely, the average initial and gel-point concentrations of reactive groups, $(c_{a0} + c_{b0})/2$ and $(c_{ac} + c_{bc})/2$, respectively. Also, c_{int} is taken as the sum of $c_{int,i}$ of eqn. 3 over all ring sizes with

$$c_{int} = \sum_{i=1}^{\infty} c_{int,i} = \text{Pab} \cdot \phi(1, 3/2) \qquad (76)$$

where $\phi(1, 3/2)$ is defined by eqn. 31. The summation implies that there is only one opportunity to form a ring structure of each size. This simplification comes from Kilb's treatment which is really limited[59] to the case $f = 3$.

The gelation data discussed in the previous section[8,61,62] on poly-urethane- and polyester-forming systems were analysed[48] according to eqns. 73 to 76 showing that the functional dependence of λ upon α_c given by

Frisch's expression was preferable to that given by Kilb's expression. Plots of λ_F versus c_{ext}^{-1} were linear for $c_{ext} = (c_{a0} + c_{b0})/2$ and showed a slight decrease in slope with dilution for $c_{ext} = (c_{ac} + c_{bc})/2$. Values of Pab for the various reaction systems were derived using the two values of c_{ext}, to give two sets of values of b for the various chain structures. It was found that b varied with chain structure in the way expected, in that a larger proportion of the more flexible oxypropylene units in the chain of v bonds gave smaller values of b. The absolute values were somewhat smaller than those expected from solution properties, this being attributed to approximations in the theories.

Recently, Ahmad and Stepto[59] have shown that the agreement with Frisch–Kilb theory, as the use of eqns. 73, 75 and 76 may be termed, was fortuitous. They removed the assumption that only small amounts of intra-molecular reaction occurred, and the Frisch expression was then shown seriously to underestimate the opportunities for intra-molecular reaction. A more rigorous development of Kilb's model for the gelation condition was given, resulting in a new gel-point expression and allowing $RA_2 + RB_f$ polymerisations to be treated. The development of this expression is now discussed.

Kilb used a linear sequence of repeating structures to define the condition for gelation. The sequence is generalised in Fig. 20 to an $RA_2 + RB_f$ polymerisation and some 'side-group' structure is specified. Assume that the sequence to the right of B^1 exists, then each branch unit (of generation i) will have $f - 2$ unreacted pendant groups. (The number can be larger than $f - 2$ if longer side chains, which must themselves contain branch units, are considered.) The unreacted pendant groups can be A and B groups, as indicated in Fig. 20. Because a path from B^1 to B^2 gives a repeating unit of the sequence, gelation occurs when the probability that the sequence

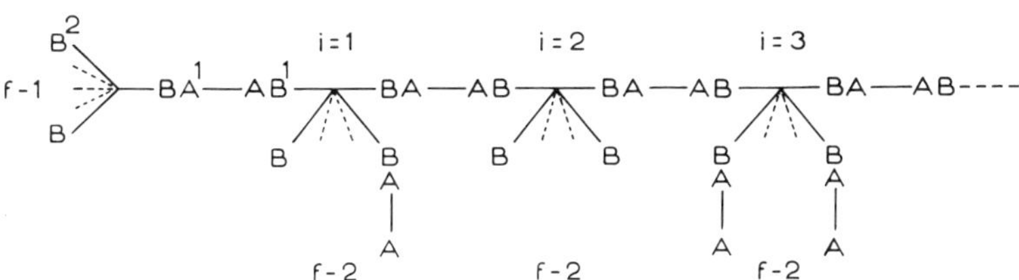

FIG. 20.   Development of Kilb's linear sequence to define the condition for gelation in an $RA_2 + RB_f$ polymerisation.[59] (Reproduced from Ahmad, Z. and Stepto, R. F. T., *Colloid Polym. Sci.*, **258** (1980), p. 663, by permission of Dr Dietrich Steinkopff Verlag, Darmstadt ©.)

continues from B^1 to B^2 is equal to unity. This probability is the product of the probabilities of two inter-molecular reaction steps, at each of which there is the possibility of intra-molecular reaction. The existence of two opportunities for intra-molecular reaction had previously been neglected.[38]

Consider the reaction of B^1. It reacts inter-molecularly with probability

$$p_b(1 - \lambda_a) \tag{77}$$

where

$$\lambda_a = \sum_{i=1}^{\infty} r_{i,a} \tag{78}$$

with $r_{i,a}$ the probability of intra-molecular reaction with an A group pendant to branch unit i.

$$r_{i,a} = \frac{c_{a,\text{int},i}}{c_{a,\text{int}} + c_{a,\text{ext}}} \tag{79}$$

where $c_{a,\text{ext}}$ is some instantaneous concentration of unreacted A groups in the reaction mixture, i.e.

$$c_{a,\text{ext}} = c_a' = c_{a0}(1 - p_a') \tag{80}$$

where p_a' must be chosen arbitrarily; $c_{a,\text{int},i}$ is the concentration, around B^1, of A groups pendant to branch unit i, and

$$c_{a,\text{int}} = \sum_{i=1}^{\infty} c_{a,\text{int},i} \tag{81}$$

The simplest expression that can be used for $c_{a,\text{int},i}$ is

$$c_{a,\text{int},i} = (f - 2) \cdot \frac{\text{Pab}}{i^{3/2}} \cdot \frac{c_a'}{c_a' + c_b'} \tag{82}$$

The factor $c_a'/(c_a' + c_b')$ on the right-hand side of eqn. 82 puts the probability that a given pendent arm ends in an A group equal to the overall mole fraction of unreacted A groups at that point in the reaction when the external concentrations of unreacted A and B groups are c_a' and c_b'. Evaluation of $r_{i,a}$ according to eqns. 79 to 82 and substitution in eqn. 78 gives

$$\lambda_a = \frac{c_{a,\text{int}}}{c_{a,\text{int}} + c_{a,\text{ext}}} = \frac{(f-2)\text{Pab} \cdot \phi(1, 3/2) \cdot c_a'/(c_a' + c_b')}{(f-2)\text{Pab} \cdot \phi(1, 3/2) \cdot c_a'/(c_a' + c_b') + c_a'} \tag{83}$$

The probability that A^1 has reacted inter-molecularly is given by the expression corresponding to expression 77,

$$p_a(1 - \lambda_b) \tag{84}$$

with

$$\lambda_b = \frac{c_{b,int}}{c_{b,int} + c_{b,ext}} = \frac{(f-2)\text{Pab}. \phi(1, 3/2). c_b'/(c_a' + c_b')}{(f-2)\text{Pab}. \phi(1, 3/2). c_b'/(c_a' + c_b') + c_b'} \tag{85}$$

Reaction of A^1 leads to $(f-1)$ possible paths for continuing the linear sequence to B^2. Thus, gelation occurs when

$$(f-1). p_b(1 - \lambda_a). p_a(1 - \lambda_b) = 1$$

giving

$$\alpha_c(f-1)(1 - \lambda_a)(1 - \lambda_b) = 1 \tag{86}$$

Further, comparison of eqns. 83 and 85 shows that $\lambda_a = \lambda_b = \lambda_{ab}$, with

$$\lambda_{ab} = \frac{(f-2)\text{Pab}. \phi(1, 3/2)}{(f-2)\text{Pab}. \phi(1, 3/2) + (c_a' + c_b')} \tag{87}$$

and

$$\alpha_c(f-1)(1 - \lambda_{ab})^2 = 1 \tag{88}$$

Comparison of eqn. 88 with eqn. 74 shows that $2\lambda_{ab} = \lambda_K$ for small λ_{ab} and $f = 3$. This is consistent with the use of $(c_{a0} + c_{b0})/2$ and $(c_{ac} + c_{bc})/2$ for c_{ext} in the previous interpretation of λ_K, as in eqn. 75. The present arguments give $c_{ext} = c_a' + c_b'$, and the $(c_{a0} + c_{b0})/2$ and $(c_{ac} + c_{bc})/2$ used previously are equivalent to $(c_a' + c_b')/2$.

Equation 88 is the simplest result derivable from the correct application of the Kilb model of a linear sequence of branch units. The two factors $(1 - \lambda_{ab})$ arise from the two opportunities for intra-molecular reaction in passing from one branch unit to the next. The general form of expression derivable from the Kilb model is given by eqn. 86. In the present case, $\lambda_a = \lambda_b$ because of the simplified structure given to the side groups. A more complete consideration of their structures would lead to $\lambda_a \neq \lambda_b$ and possibly to expressions which could be applied discriminately to gelation data[61] from reactions with unequal concentrations of A and B groups $(r \neq 1)$. Accepting the simplified structures, the theory allows $f - 2$ reactive ends per branch unit and is, therefore, not limited to the case $f = 3$.

Equations 87 and 88 lead to the equations

$$\alpha_c^{\frac{1}{2}}(f-1)^{\frac{1}{2}} - 1 = \frac{\lambda_{ab}}{1-\lambda_{ab}} = \lambda'_{ab} \qquad (89)$$

and

$$\lambda'_{ab} = \frac{c_{int}}{c_{ext}} = \frac{(f-2)\mathrm{Pab}.\,\phi(1,3/2)}{c_{ext}} \qquad (90)$$

Now c_{ext} can be put equal to $(c_{a0} + c_{b0})$ and $(c_{ac} + c_{bc})$ as extreme values. Plots of λ'_{ab} versus $(c_{a0} + c_{b0})^{-1}$ and $(c_{ac} + c_{bc})^{-1}$, the initial and gel-point dilutions, are shown in Figs. 21 and 22 using the data of Smith and Stepto[62] from Fig. 19. The plots show approximate linearity when $(c_{a0} + c_{b0})^{-1}$ is used as abscissa and decreasing slopes when $(c_{ac} + c_{bc})^{-1}$ is used. Similar behaviour is observed[110] if other published data (refs. 8, 48, 61) are analysed according to eqns. 89 and 90.

The plots show in alternative fashion the increase in intra-molecular reaction (λ'_{ab}) with increase in dilution and decrease in v, apparent in Figs. 18 and 19. The slopes or initial slopes can be analysed[59] according to eqns. 4 and 90 and the values of b so derived are given in Table 6. Chain stiffness would be expected to increase with v_{ac}/v (the fractional length of acid chloride residue in the chain forming the smallest ring) and this trend is observed. Thus, although intra-molecular reaction is characterised principally by v, the secondary dependence on chain structure noted

TABLE 6

VALUES OF b DERIVED[59] FROM FIGURES 21 AND 22 USING EQNS. 4, 89 AND 90

System	v	v_{ac}/v^{a}	b (nm) $(i)^{b}$	b (nm) $(ii)^{c}$
2. Sebacoyl chloride/LHT240	41	0·268	0·318	0·508
1. Adipoyl chloride/LHT240	37	0·189	0·313	0·480
4. Sebacoyl chloride/LHT112	70	0·157	0·293	0·433
3. Adipoyl chloride/LHT112	66	0·106	0·270	0·399
6. Sebacoyl chloride/LG56	136	0·081	0·267	0·390
5. Adipoyl chloride/LG56	132	0·053	0·260	0·371

^{a} v_{ac}/v = fractional length of acid chloride residue in chain forming smallest ring.
^{b} (i) $c_{ext} = (c_{a0} + c_{b0})$.
^{c} (ii) $c_{ext} = (c_{ac} + c_{bc})$.
(Reproduced in part from Ahmad, Z. and Stepto, R. F. T., *Colloid Polym. Sci.*, **258** (1980), p. 668, by permission of Dr Dietrich Steinkopff Verlag, Darmstadt ©.)

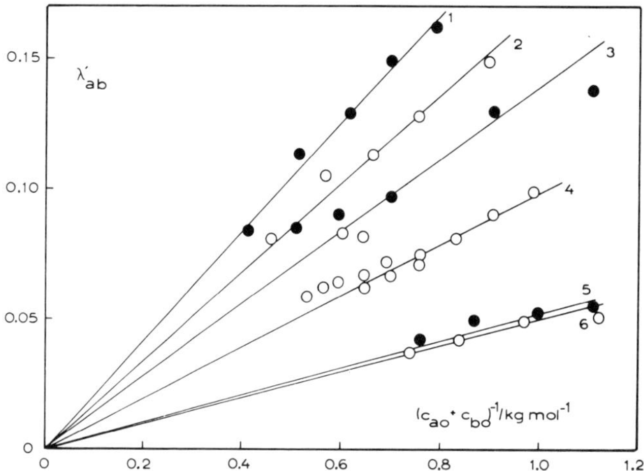

FIG. 21.   Ring-forming parameter (λ'_{ab}) versus initial dilution of reactive groups $((c_{a0} + c_{b0})^{-1})$. Analysis[59] of gelation data in Fig. 19 according to eqns. 89 and 90 with $c_{ext} = c_{a0} + c_{b0}$. Systems 1–6 as for Fig. 19. (Reproduced from Ahmad, Z. and Stepto, R. F. T., *Colloid Polym. Sci.*, **258** (1980), p. 663, by permission of Dr Dietrich Steinkopff Verlag, Darmstadt ©.)

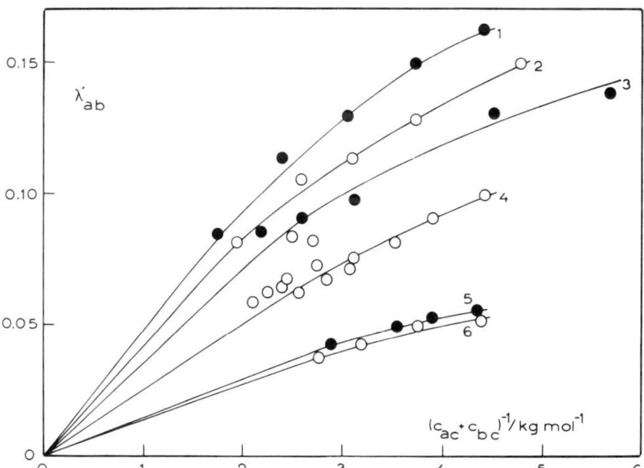

FIG. 22.   Ring-forming parameter (λ'_{ab}) versus gel-point dilution of reactive groups $((c_{ac} + c_{bc})^{-1})$. Analysis[59] of gelation data in Fig. 19 according to eqns. 89 and 90 with $c_{ext} = c_{ac} + c_{bc}$. Systems as for Fig. 19. (Reproduced from Ahmad, Z. and Stepto, R. F. T., *Colloid Polym. Sci.*, **258** (1980), p. 663, by permission of Dr Dietrich Steinkopff Verlag, Darmstadt ©.)

previously for linear systems (see Table 5 and following discussion) is apparent. More intra-molecular reaction occurs with the more flexible chains (smaller values of b).

A detailed discussion of the absolute values of b in Table 6 has been given.[59] It would be expected that the values in column (ii) would be more comparable with those derived from unperturbed, linear chains. Relatively more intra-molecular reaction occurs near the gel-point so that $(c_{ac} + c_{bc})$ is the better approximation to the average value required for c_{ext}. The values in column (ii) are in fact somewhat higher than would be expected and this may be attributed to the fact that the model of chain growth (Fig. 20) probably overestimates the opportunities for intra-molecular reaction.[111,112]

In conclusion, the assumptions used in approximate theories of gelation may be listed:[59]

(i)   The many isomeric structures which define opportunities for intra-molecular reaction may be simplified;

(ii) c_{ext} is independent of extents of reaction;

(iii)   Gaussian chain statistics may be used to define c_{int} in branched molecules;

(iv) $c_{int} \ll c_{ext}$;

(v)   The instantaneous probability of intra-molecular reaction is given by a ratio of concentrations, for example $r_i = c_{int,i}/(c_{int} + c_{ext})$.

Assumptions (i), (iii) and (v) are also being employed in the more complete theories describing the inter- and intra-molecular growth of molecular species during non-linear random polymerisations (e.g. refs. 20, 50, 81).

The work of Ahmad and Stepto[59] has shown that assumption (iv) is unnecessary. Given assumptions (i) and (ii), it is doubtful whether it is worth attempting to improve on assumption (iii). Deviations from Gaussian statistics may be expected if the number of bonds in the chain forming the smallest ring structure, v, is small, or if the chains are expanded (see section 3). Assumption (ii) is essential if a simple, analytical expression for the condition for gelation, characterised by a single ring-forming parameter, is to be achieved. However, it does mean that the value of c_{ext} used must be chosen arbitrarily, and the functional form of the dependence of λ'_{ab} on c_{ext}^{-1} cannot be predicted. Thus, the linear plots of Fig. 21 compared with the curved plots of Fig. 22 do not mean that $(c_{a0} + c_{b0})$ is a better choice than $(c_{ac} + c_{bc})$ for c_{ext}. This lack of certainty of the functional form of the dependence of λ'_{ab} on dilution has serious implications regarding the use[41,79,80] of approximate gel-point expressions for the determination of reactant functionalities. Finally, assumption (v) is related to the connection between detailed reaction mechanisms and the random reaction of functional groups (see section 3 and refs. 51 and 59).

10. SUMMARY

It is clear that intra-molecular reaction is an integral part of the random polymerisation process. In irreversible polymerisations the extent to which it occurs depends on dilution (c_{ext}), reactant functionalities (f), molar mass (v), and chain flexibility (b). Even for quite small values of v, intra-molecular reaction can probably be neglected for linear, bulk polymerisations (Fig. 6), except possibly for $p \cong 1$. For gelling systems, much more pre-gel intra-molecular reaction occurs than would occur in a linear system with the same value of v (Fig. 16), and the properties of the network materials formed can be markedly affected by such intra-molecular reaction.[63-65]

Studies of linear ring-chain equilibria are important in their own right and they also provide a basis for the understanding of the factors affecting ring formation in irreversible polymerisations. The interpretation of intra-molecular reaction in linear, irreversible polymerisations is reasonably complete, given the data presently available (total ring fraction data compared with distributions of ring sizes for systems at equilibrium). Intra-molecular reaction in non-linear polymerisations and its effect on gel-points and network properties are less well characterised. Although much theoretical work has been carried out in this area, more data linking these three aspects and using a variety of model systems are needed. Such data could provide a firm basis for the examination of the more quantitative theories of random polymerisation and lead to the prediction of materials properties for given reactants and reaction conditions.

REFERENCES

1. Flory, P. J., *Principles of Polymer Chemistry*, Ithaca, New York, Cornell University Press, 1953, (*a*) Chapter 2, (*b*) Chapter 8, (*c*) Chapter 9.
2. Lenz, R. W., *Organic Chemistry of Synthetic High Polymers*, New York, Interscience Publishers, 1967, Chapter 1.
3. Carothers, W. H., *J. Amer. Chem. Soc.*, **51** (1929), p. 2548.
4. Stanford, J. L. and Stepto, R. F. T., *J. Chem. Soc., Faraday Trans. I*, **71** (1975), p. 1292.

 5. FLORY, P. J., *Chem. Revs.*, **39** (1946), p. 137.
 6. STOCKMAYER, W. H., *J. Chem. Phys.*, **11** (1943), p. 45; **12** (1944), p. 125.
 7. STOCKMAYER, W. H., *J. Polym. Sci.*, **9** (1952), p. 69; **11** (1953), p. 424.
 8. HOPKINS, W., PETERS, R. H. and STEPTO, R. F. T., *Polymer*, **15** (1974), p. 315.
 9. HOEVE, C. A. J., *J. Polym. Sci.*, **21** (1956), pp. 1, 11.
10. CASE, L. C., *J. Polym. Sci.*, **26** (1957), p. 333.
11. FUKUI, K. and YAMABE, T., *J. Polym. Sci.*, **45** (1960), p. 305.
12. DIGIACOMO, A., *J. Polym. Sci.*, **47** (1960), p. 435.
13. JONASON, M., *J. Appl. Polym. Sci.*, **4** (1960), p. 129.
14. KAHN, A., *J. Polym. Sci.*, **49** (1961), p. 283.
15. LILLEY, H. S., *J. Appl. Polym. Sci.*, **5** (1961), p. S16.
16. GORDON, M. and SCANTLEBURY, G. R., *Trans. Faraday Soc.*, **60** (1964), p. 604.
17. YAMABE, T. and FUKUI, K., *Bull. Chem. Soc. Japan*, **37** (1964), p. 1061.
18. TANAKA, Y. and KAKUICHI, H., *J. Polym. Sci. Part A*, **3** (1965), p. 3279.
19. GORDON, M. and SCANTLEBURY, G. R., *Proc. Roy. Soc. (London)*, **A292** (1966), p. 380.
20. GORDON, M. and SCANTLEBURY, G. R., *J. Chem. Soc. B* (1967), p. 1.
21. GORDON, M. and SCANTLEBURY, G. R., *J. Polym. Sci. Part C*, **16** (1968), p. 3933.
22. DUŠEK, K. and PRINS, W., *Adv. Polym. Sci.*, **6** (1969), p. 1.
23. FOGIEL, A. W. and STEWART, C. W., SR, *J. Polym. Sci.*, *Part A-2*, **7** (1969), p. 1116.
24. GORDON, M. and PARKER, T. G., *Proc. Roy. Soc. Edin. A*, **69** (1970), p. 13.
25. FRENCH, D. M., *J. Macromol. Sci.*, *Chem.*, **A5** (1971), p. 1123.
26. KONSHTEIN, S. E. and PIS'MEN, L. M., *Dokl. Akad. Nauk. SSSR*, **196** (1971), p. 858.
27. PIS'MEN, L. M. and KUCHANOV, S. J., *Vysokomol. Soyed.*, *Ser. A*, **13** (1971), p. 791.
28. URAGAMI, T. and OIWA, M., *Makromol. Chem.*, **153** (1972), p. 255.
29. ZIEGEL, K. D., FOGIEL, A. W. and PARISER, R., *Macromolecules*, **5** (1972), p. 95.
30. COHEN, C., GIBBS, J. H. and FLEMING, P. D., III, *J. Chem. Phys.*, **59** (1973), p. 5511.
31. DUŠEK, K., *J. Polym. Sci.*, *Phys. Ed.*, **12** (1974), p. 1089.
32. DUŠEK, K., ILAVSKÝ, M. and LUŇÁK, S., *J. Polym. Sci.*, *Polymer Symp*, **53** (1975), p. 29; LUŇÁK, S. and DUŠEK, K., *J. Polym. Sci.*, *Polymer Symp.*, **53** (1975), p. 45.
33. KODAMA, Y., TEMPLE, W. B. and ROSS-MURPHY, S. B., *Brit. Polym. J.*, **9** (1977), p. 117.
34. DURAND, D. and BRUNEAU, C. M., *J. Polym. Sci.*, *Part A-2*, **17** (1979), pp. 273, 295.
35. JACOBSON, H. and STOCKMAYER, W. H., *J. Chem. Phys.*, **18** (1950), p. 1600.
36. FRISCH, H. L., paper presented to 128th meeting Amer. Chem. Soc., Polymer Division, Minneapolis (1955).
37. HARRIS, F. E., *J. Chem. Phys.*, **23** (1955), p. 1518.
38. KILB, R. W., *J. Phys. Chem.*, **62** (1958), p. 969.
39. CARMICHAEL, J. B. and HEFFEL, J., *J. Phys. Chem.*, **69** (1965), p. 2213.
40. FLORY, P. J. and SEMLYEN, J. A., *J. Amer. Chem. Soc.*, **88** (1966), p. 3209.

41. FOGIEL, A. W., *Macromolecules*, **2** (1969), p. 581.
42. MORAWETZ, H. and GOODMAN, N., *Macromolecules*, **3** (1970), p. 699.
43. IRZHAK, V. J., KUZUB, L. J. and ENIKOLOPYAN, N. S., *Dokl. Akad. Nauk. SSSR*, **201** (1971), p. 1382.
44. STEPTO, R. F. T. and WAYWELL, D. R., *Makromol. Chem.*, **152** (1972), p. 263.
45. GORDON, M. and TEMPLE, W. B., *Makromol. Chem.*, **152** (1972), p. 277.
46. GORDON, M. and TEMPLE, W. B., *Makromol. Chem.*, **160** (1972), p. 263.
47. TEMPLE, W. B., *Makromol. Chem.*, **160** (1972), p. 277.
48. STEPTO, R. F. T., *Faraday Disc. Chem. Soc.*, **57** (1974), p. 69.
49. GORDON, M. and ROSS-MURPHY, S. B., *Pure Appl. Chem.*, **43** (1975), p. 1.
50. ROSS-MURPHY, S. B., *J. Polym. Sci.,|Polymer Symp.*, **53** (1975), p. 11.
51. STANFORD, J. L., STEPTO, R. F. T. and WAYWELL, D. R., *J. Chem. Soc. Faraday Trans. I*, **71** (1975), p. 1308.
52. ALLEN, G., *Proc. Roy. Soc. (London)*, **A351** (1976), p. 381.
53. (*a*) FLORY, P. J., SUTER, U. W. and MUTTER, M., *J. Amer. Chem. Soc.*, **98** (1976), p. 5733; (*b*) SUTER, U. W., MUTTER, M. and FLORY, P. J., *J. Amer. Chem. Soc.*, **98** (1976), p. 5740; (*c*) MUTTER, M., SUTER, U. W. and FLORY, P. J., *J. Amer. Chem. Soc.*, **98** (1976), p. 5745.
54. SEMLYEN, J. A., *Adv. Polym. Sci.*, **21** (1976), p. 41.
55. DUŠEK, K. and VOJTA, V., *Brit. Polym. J.*, **9** (1977), p. 164.
56. DUŠEK, K., MADHOUD, M. and ILAVSKÝ, M., *Brit. Polym. J.*, **9** (1977), p. 172.
57. MUTTER, M., *J. Amer. Chem. Soc.*, **99** (1977), p. 8307.
58. DUŠEK, K., GORDON, M. and ROSS-MURPHY, S. B., *Macromolecules*, **11** (1978), p. 236.
59. AHMAD, Z. and STEPTO, R. F. T., *Colloid Polym. Sci.*, **258** (1980), p. 663.
60. STANFORD, J. L. and STEPTO, R. F. T., *Brit. Polym. J.*, **9** (1977), p. 124.
61. PETERS, R. H. and STEPTO, R. F. T., in *The Chemistry of Polymerisation Processes*, London, Soc. Chem. Ind. Monograph No. 20, 1965, p. 157.
62. SMITH, R. S. and STEPTO, R. F. T., *Makromol. Chem.*, **175** (1974), p. 2365.
63. STEPTO, R. F. T., *Polymer*, **20** (1979), p. 1324.
64. HUNT, N. G. K., STEPTO, R. F. T. and STILL, R. H., *Preprints IUPAC 26th Intl. Symposium on Macromolecules, Mainz, 1979*, p. 697; results reported in part in ref. 63.
65. FAṢINA, A. B. and STEPTO, R. F. T., *Makromol. Chem.* (1981), in press; results reported in part in *Preprints IUPAC Discussion Conference on Polymer Networks, Karlovy Vary, 1980*, p. C49–1.
66. FLORY, P. J., *Statistical Mechanics of Chain Molecules*, New York, Interscience Publishers, 1969.
67. (*a*) DODGSON, K. and SEMLYEN, J. A., *Polymer* **18** (1977), p. 1265; (*b*) DODGSON, K., SYMPON, D. and SEMLYEN, J. A., *Polymer*, **19** (1978), p. 1285; (*c*) HIGGINS, J. S., DODGSON, K. and SEMLYEN, J. A., *Polymer*, **20** (1979), p. 553; (*d*) DODGSON, K., BANNISTER, D. J. and SEMLYEN, J. A., *Polymer*, **21** (1980), p. 663; (*e*) EDWARDS, C. J. C., STEPTO, R. F. T. and SEMLYEN, J. A., *Polymer*, **21** (1980), p. 781.
68. KIENLE, R. H., VAN DER MEULEN, P. A. and PETKE, F. E., *J. Amer. Chem. Soc.*, **61** (1939), pp. 2258, 2268; KIENLE, R. H. and PETKE, F. E., *J. Amer. Chem. Soc.*, **62** (1940), p. 1053; **63** (1941), p. 481.
69. FLORY, P. J., *J. Amer. Chem. Soc.*, **63** (1941), p. 3083.

70. STOCKMAYER, W. H. and WEIL, L. L., results quoted by STOCKMAYER, W. H. in *Advancing Fronts in Chemistry*, Twiss, S. B. (Ed.), New York, Reinhold Publishing Corp., 1945, Chapter 6.
71. PRICE, F. P., GIBBS, J. H. and ZIMM, B. H., *J. Phys. Chem.*, **62** (1958), p. 972.
72. PRICE, F. P., *J. Phys. Chem.*, **62** (1958), p. 977.
73. NAKAMURA, Y., KOIDE, J. and NEGISHI, M., *J. Polym. Sci.*, **42** (1960), p. 203.
74. NEGISHI, M., NAKAMURA, Y. and KOIDE, J., *J. Polym. Sci.*, **43** (1960), p. 241.
75. BATES, R. F. and HOWARD, G. J., *J. Polym. Sci. Part C*, **16** (1968), p. 921.
76. BERNARDO, J. J. and BRUINS, P., *J. Paint Technol.*, **40** (1968), p. 558.
77. STRECKER, R. A. H. and FRENCH, D. M., *J. Appl. Polym. Sci.*, **12** (1968), p. 1697.
78. GORDON, M., TEMPLE, W. B., PARKER, T. G. and LOVE, J. A., *J. prakt. Chem.*, **313** (1971), p. 411.
79. GUISE, G. B. and ZOCH, C. G., *J. Macromol. Sci., Chem.*, **A10** (1976), p. 1161.
80. GUISE, G. B. and ZOCH, C. G., *J. Polym. Sci., Polym. Symp.*, **55** (1976), p. 61.
81. CAWSE, J. L., STANFORD, J. L. and Stepto, R. F. T., *Preprints IUPAC 26th Intl. Symposium on Macromolecules, Mainz, 1979*, p. 693.
82. KUHN, W., *Kolloid-Z.*, **68** (1934), p. 2.
83. HAWARD, R. N., *Trans. Faraday Soc.*, **46** (1950), p. 204.
84. BROWN, J. F. and SLUSARCZUK, G. M. J., *J. Amer. Chem. Soc.*, **87** (1965), p. 931.
85. WRIGHT, P. V., *J. Polym. Sci., Polymer Phys.*, **11** (1973), p. 51.
86. DOSTAL, H. and RAFF, R., *Z. physik. Chem.*, **1332** (1936), p. 117; *Monatsh*, **68** (1936), p. 188.
87. TANFORD, C., *Physical Chemistry of Macromolecules*, New York, John Wiley & Sons, 1961, Chapter 9.
88. CHOJNOWSKI, J., ŚCIBIOREK, M. and KOWALSKI, J., *Makromol. Chem.*, **178** (1977), p. 1351.
89. STEPTO, R. F. T. and WAYWELL, D. R., *Makromol. Chem.*, **152** (1972), p. 247.
90. GREENSHIELDS, J. N., PETERS, R. H. and STEPTO, R. F. T., *J. Chem. Soc.* (1964), p. 5101.
91. HOPKINS, W., Ph.D. Thesis, University of Manchester, UK, 1967.
92. GORDON, M., *Proc. Roy. Soc. (London)*, **A268** (1962), p. 240.
93. DOBSON, G. R. and GORDON, M., *J. Chem. Phys.*, **41** (1964), p. 2389; **43** (1965), p. 705.
94. GORDON, M. and MALCOLM, G. N., *Proc. Roy. Soc. (London)*, **A295** (1966), p. 29.
95. IMAI, S. and GORDON, M., *J. Chem. Phys.*, **50** (1969), p. 3889.
96. KAJIWARA, K., *J. Chem. Phys.*, **54** (1971), p. 296; *Polymer*, **12** (1971), p. 57.
97. SMALL, P. A., *Polymer*, **13** (1972), p. 536.
98. KAJIWARA, K. and GORDON, M., *J. Chem. Phys.*, **59** (1973), p. 3623.
99. BURCHARD, W., KAJIWARA, K., GORDON, M., KÁLAL, J. and KENNEDY, J. W., *Macromolecules*, **6** (1973), p. 642.
100. PENICHE-COVAS, C. A. L., DEV, S. B., GORDON, M., JUDD, M. and KAJIWARA, K., *Faraday Disc. Chem. Soc.*, **57** (1974), p. 165.
101. MOORE, W. R., *An Introduction to Polymer Chemistry*, London, University of London Press, 1963., p. 26.

102. COWIE, J. M. G., *Polymers: Chemistry and Physics of Modern Materials*, Aylesbury, Intertext Books, 1973, p. 42.
103. CAROTHERS, W. H., *Trans. Faraday Soc.*, **32** (1936), p. 39.
104. BRUNEAU, C. M., *Compt. Rend. Acad. Sci. (Paris)*, **C264** (1967), pp. 758, 1168, 1422.
105. WHITEWAY, S. G., SMITH, J. B. and MASSON, C. R., *Canad. J. Chem.*, **48** (1970), pp. 33, 201, 1432, 1456.
106. GORDON, M. and JUDD, M., *Nature (London)*, **234** (1971), p. 96.
107. MASSON, C. R., SMITH, J. B. and WHITEWAY, S. G., *Nature (London)*, **234** (1971), p. 97.
108. STEPTO, R. F. T., *Canad. J. Chem.*, **52** (1974), p. 1188.
109. MASSON, C. R. and WHITEWAY, S. G., *Canad. J. Chem.*, **52** (1974), p. 1190.
110. AHMAD, Z., Ph.D. Thesis, University of Manchester, UK, 1978.
111. FLORY, P. J., *Faraday Disc. Chem. Soc.*, **57** (1974), p. 87.
112. STEPTO, R. F. T., *Faraday Disc. Chem. Soc.*, **57** (1974), p. 91.

Chapter 4

NETWORK FORMATION BY CHAIN CROSS-LINKING (CO)POLYMERISATION†

KAREL DUŠEK

*Institute of Macromolecular Chemistry,
Czechoslovak Academy of Sciences, Czechoslovakia*

SUMMARY

The present review stresses the importance of cyclisation in free-radical (co)polymerisation of monovinyl and polyvinyl monomers. The high extent of cyclisation is determined by the nature of the chain process and strongly influences the copolymerisation behaviour and structure of products. Only at a very low concentration of the polyvinyl monomer, the modified copolymerisation and cross-linking theories can conform with experiments, but at medium and high concentrations of the cross-linker, microgel-like particles are formed from the very beginning of the reaction. These microgel-like particles are joined together by their peripheral double bonds only to form later an infinite (but often heterogeneous) network.

In this review, ring-free theories and theories allowing for perturbation by cyclisation for copolymerisation and networks formation are compared with experiments pertaining to the pre-gel and post-gel course of reaction and structure of the products. Effects of the special features of the polymerisation process coupled with phase separation on the formation and structure of heterogeneous (macroporous) networks serving as sorbents or sorbent matrices in many applications are also discussed.

† In this chapter (co)polymerisation refers to copolymerisation as well as polymerisation, i.e. homopolymerisation. Copolymerisation (without brackets) refers exclusively to copolymerisation.

143

SYMBOLS AND ABBREVIATIONS

a	fraction of cross-linked units engaged in inter-molecular bonds
A	front factor in the theory of rubber elasticity
b	number of atoms which a monomer unit contributes to the main chain
c_{DV}	concentration of the divinyl monomer
C_0, C_p	molar concentrations of all and reacted double bonds, respectively
e	proportionality constant in the trapped entanglement contribution to the equilibrium modulus
f_1	$= M_1/M_2$, $f_2 = 1 - f_1$ in binary copolymerisation
F_1	$= m_1/m_2$
g_0	$= \langle s_0^2 \rangle_b / \langle s_0^2 \rangle_1$
g'	$= [\eta]_b / [\eta]_1$
G_e	equilibrium shear modulus
G_{ec}, G_{ee}	chemical and trapped entanglement contributions to G_e
k_{ij}	rate constant for reaction of the free radical of type i with double bond of type j
l, l_s	length of a segment or a statistical segment
m_i	molar concentration of units i (reacted vinyls of type i) in the copolymer
m_ρ	molar concentration of cross-linked units
m_x	number of monomer units in the generation x
M_i	molar concentration of vinyls of type i
M, M_n, M_w	molecular weight and its number- and weight-averages
n	number of repeat units
$n_{P,\rho}$	number fraction of primary chains composed of P units the fraction ρ of them being cross-linked
n_s	number of statistical segments in a chain
N_A	Avogadro's number
p	fraction of divinyl units bearing a pendant double bond $(p + a + s = 1)$
p_d	dry porosity
P, P_n, P_w	degree of polymerisation and its number- and weight-averages
P^0	degree of polymerisation of primary chains
P_c, P_{cx}	cyclisation probabilities
Q	number of main chain atoms in a statistical segment

r_i, r_{ij}	copolymerisation parameters
R	separation distance of chain ends
s	fraction of cross-linked units engaged in cycles $(p + a + s = 1)$
$\langle s^2 \rangle$	mean-square radius of gyration
t	time
T_{eg}	trapping factor for the gel in the theory of trapped entanglements
v_0	volume fraction of monomers in the initial mixture; $1 - v_0 =$ volume fraction of the diluent
V	volume
$w_{2,0}$	initial weight fraction of the cross-linker
$W(R)$	end-to-end distance distribution function
x	unsaturation (fraction of unreacted double bonds related to the content of double bond in the bis-unsaturated monomer)
x_1, x_2, x_3	molar fractions of the monovinyl monomer, divinyl monomer with pendant double bond and doubly reacted divinyl monomer, respectively; $x_1 = m_1/(m_1 + m_2 + m_3/2)$, $x_2 = m_2/(m_1 + m_2 + m_3/2)$, $x_3 = (m_3/2)/(m_1 + m_2 + m_3/2)$
x_{DV}^0	initial molar fraction of the cross-linker
Y	$= P_c m_2'/C_0$ cyclisation parameter
α	$= a/(a + s)$, efficiency of cross-linking
ε	fraction of effective cross-links
$[\eta]_l, [\eta]_b$	intrinsic viscosities of linear and branched polymers, respectively
v	concentration of cross-linked units
v_e	concentration of elastically active network chains (EANC)
v_{ec}	chemical contribution to v_e
ξ	molar conversion of double bonds
ξ_c	critical molar conversion of double bonds at the gel-point
ξ_w	weight conversion of monomer to polymer
ρ	fraction of cross-linked units in chains
ρ_{id}	ideal fraction of cross-linked units = fraction of cross-linked units engaged in inter-molecular bonds = $s/(a + s)$
σ	fraction of bonds engaged in cycles

φ_i	molar fractions of vinyls of type i
φ_i^0	initial value of φ_i
Λ	cyclisation factor (eqn. 27)
Φ	Flory constant (eqn. 52)
Φ_i	molar fraction of reacted vinyls of type i
Ω	average number of unreacted functional groups on a unit that has already reacted
DAP, DAI, DAT	diallyl ortho-, iso- and tere-phthalate
DVB	divinylbenzene
EDMA	ethylene dimethacrylate
MBA	methylene-bisacrylamide
MMA	methyl methacrylate
PMMA	poly(methyl methacrylate)
PETMA	pentaerythritol tetramethacrylate
PST	polystyrene
ST	styrene

1. INTRODUCTION

In polymerisation reactions, network formation is possible if at least one of the initial components has a functionality higher than two. According to the specificity of chemical mechanism, the reactions of network formation may be divided into three groups:

(a)   stepwise polyaddition or polycondensation,
(b)   cross-linking (vulcanisation) of existing (primary) chains,
(c)   cross-linking chain (co)polymerisation.

The reactions included in group (a) are characterised by their stepwise character and relatively low functionality of the components; the molecular weight of the components is usually low, but network formation from telechelic polymers via their end groups also falls within this category. A characteristic feature of such processes is a continuous change in the molecular weight distribution and a relatively high conversion of functional groups at the gel-point.

The vulcanisation reactions (group (b)) can also be considered stepwise, but the functionality of the initial components is rather high and usually identical with the degree of polymerisation of the primary chain; therefore, the degree of conversion at the gel-point is usually low. An initial polydispersity in the degree of polymerisation and thus in the functionality is very frequent.

Formation of only few polyfunctional macromolecules dissolved in the monomer is a special feature of the early stage of the chain (co)polymerisation of polyunsaturated monomers (group (c)). Because of the usually high degree of polymerisation of the formed chains, the gel-point conversion is relatively low except for the case of very low concentrations of the polyunsaturated monomer in copolymerisation with a monounsaturated one.

Although qualitatively the same characteristics accompany network formation in different systems—increase of molecular weights, broadening of the molecular weight distribution, passing through the gel-point and gradual transformation of the soluble part (sol) into the insoluble part (gel)—large differences also exist, not only in the location of the gel-point, but also in the steepness of the sol–gel transition. Any quantitative theory offering a correlation between the parameters of the initial components and progress of the reaction on the one hand, and structural parameters of the cross-linked system on the other, must take into consideration the chemical mechanism of network formation (rules of making and, possibly, breaking bonds).

An important feature in which the cross-linking reactions differ is the tendency to cyclisation. Cycles are closed by intra-molecular bonds; a bond is considered intra-molecular if it has been formed between two reactive sites already connected by at least one path of chemical bonds. Such a definition is meaningful only before the gel-point, since in the gel an overwhelming majority of bonds close cycles (closed circuits). In spite of that, one usually talks about cycles also beyond the gel-point, but then only about such cycles that do not contribute to elasticity, swelling, etc. Therefore, only those bonds are considered intra-molecular that close *elastically inactive cycles.*[1]

The tendency to cyclisation for the three types of network formation visualised in Fig. 1 differs considerably, if the situation at the beginning of the reaction is considered. In the stepwise reaction of small molecules, cycles cannot be usually formed within the same monomer molecule (either in homopolyreaction, because the cycle would be too small and too strained, or in the alternating type of polyreaction, where reaction between two groups of the same type cannot occur). The fraction of bonds engaged in cycles $\sigma_0 = s_0/(s_0 + a_0)$ (a_0 and s_0 are fractions of inter- and intra-molecular bonds, respectively) is zero, because the first bonds formed can only be of inter-molecular type. In the cross-linking of primary chains, a

                    KAREL DUŠEK

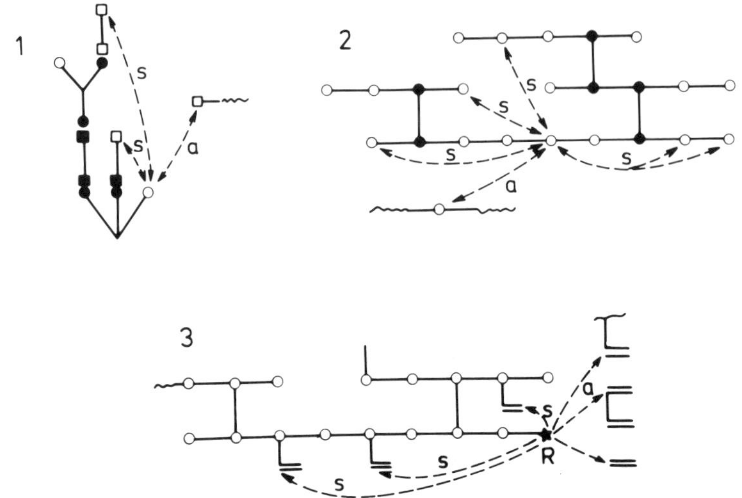

FIG. 1.   Tendency to cyclisation in different types of cross-linking reactions. (1) Step poly-addition, (2) cross-linking of primary chains, (3) cross-linking chain polymerisation.

randomly selected functionality has always a finite chance to react either inter- or intra-molecularly and σ_0 is therefore finite (for most elastomers it is of the order 10^{-1}).

In the cross-linking chain (co)polymerisation, the formed chains are initially infinitely diluted by the monomer so that an attack of a pendant double bond by a free radical of another growing chain occurs in this limit with probability zero. On the contrary, the reaction of the radical with pendant double bonds of the same chain proceeds with a non-zero probability so that $\sigma_0 = 1$. Therefore, a strong cyclisation is characteristic for the (co)polymerisation of bis-unsaturated monomers, especially in the early stages of the reaction; the reason for σ_0 approaching unity is the absence of the inter-molecular reaction. Only later on, when macro-molecules in the system become more numerous, the probability of inter-molecular reaction increases and the fraction σ falls below unity. The expected variations of the relative cyclisation intensity, σ, for the three types of cross-linking reaction are given in Fig. 2.

It is our intention to stress in this review the experimental evidence accumulated during the last decades which shows that cyclisation plays a dominant role in network formation by chain (co)polymerisation and that it affects the structure and several properties of the products formed. Only

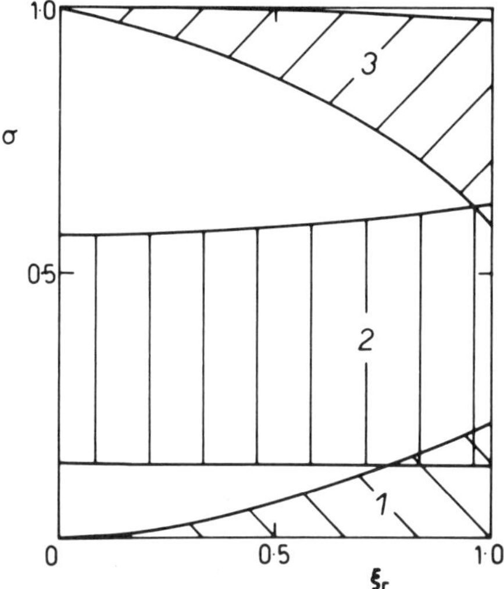

FIG. 2. Estimated fractions of cross-links wasted in cycles, σ, as a function of conversion relative to the gel-point $\xi_r = \xi/\xi_c$. (1) Step polyaddition (poly-condensation), (2) cross-linking of existing chains (vulcanisation), (3) cross-linking chain (co)polymerisation.

at very low concentrations of the polyunsaturated (e.g. divinyl) monomer in a mixture with a mono-unsaturated one, the structure of the products formed may approach the classical picture of a branching tree underlying the network formation theories of Flory and Stockmayer.[2,3] However, cyclisation can never be ruled out and the limiting law $\sigma_0 \to 1$ is generally valid, but, in contrast to the situation with extensive cyclisation, the network formation could be described using a perturbed tree-like approach (spanning tree, cf. for example ref. 4). There exists a special case of chain polymerisation in which (for steric reasons) cyclisation within the bis-unsaturated monomer unit predominates and in which essentially linear chains can be formed despite the monomer functionality of four. Such polymerisation is called cyclopolymerisation (cf., for example ref. 5, or Chapter 1 in this book). This type of chain polymerisation will not be treated here. Likewise, we will not discuss here the rate of (co)polymerisa-tion processes, but will limit ourselves to an examination of the composition and structure of (co)polymers.

At the moment there does not seem to exist any adequate theory of network formation by chain (co)polymerisation that would take into account extensive cyclisation. However, the homo- and copolymerisation of polyunsaturated monomers in the ring-free case has been dealt with in a number of papers. Although such treatments cannot be generally applied directly to real systems, they are useful in determining the effect of cyclisation and can also be applied to experimental data properly pretreated in order to eliminate the effect of cyclisation.

Therefore, we will start with the treatment of the ring-free copolymerisation of mono- and bis-unsaturated monomers, examine the possibility of inclusion of cyclisation reactions and later discuss the experimental results with respect to deviations from the expected behaviour.

2.   COPOLYMERISATION

As has been stated in section 1, the copolymerisation parameters of a bis-unsaturated and a mono-unsaturated monomer, especially, have been determined by the ring-free approximation. Several attempts have been made to treat the copolymerisation with inclusion of the cyclisation reaction, but the intensity of cyclisation is usually determined by a single cyclisation constant independent of composition and conversion. A few treatments are based on the spanning-tree approximation to the tree-like models (theory of branching processes), but none of these methods takes into account correlations due to the existing cycles. These correlations not only affect the probability of ring formation, but also violate the assumption of equal accessibility of all unreacted double bonds in the *inter-*molecular reaction steps.

2.1. Ring-free Copolymerisation

As early as in 1952, the ring-free copolymerisation of vinyl and divinyl compounds was extensively discussed in relation to vinyl–vinyl copolymerisation[6] and it was shown that the usual types of copolymerisation rate equations could be applied to each type of vinyl group and the corresponding free radical.

The reactivity of the vinyl group in the monovinyl monomer may be different from that of vinyl in the divinyl monomer; in the divinyl monomer (DV) itself the reactivities of vinyls may be equal (symmetric monomer) or different (asymmetric) monomer and the reactivities of these vinyls may be independent or dependent, i.e. the remaining vinyl may change its reactivity after one of the vinyls has reacted:

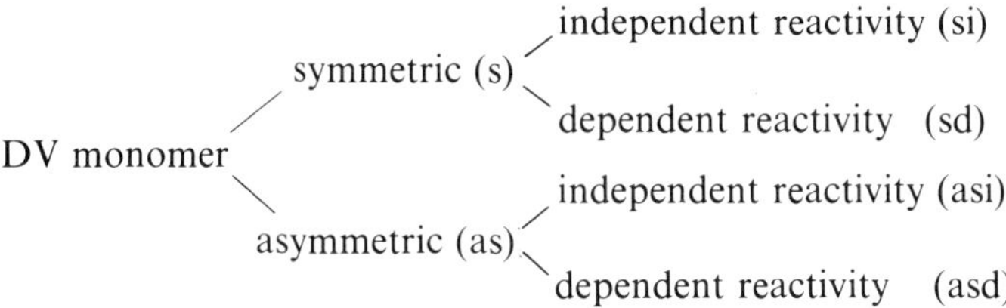

Ethylene dimethacrylate (si), *p*-divinylbenzene (sd), allyl methacrylate (asi) and some conjugated dienes (asd) can serve as examples.

The treatment of systems with independent vinyl reactivities can be reduced to simple binary or ternary copolymerisation systems only bearing in mind that, by incorporation of a divinyl molecule into the polymer, the monomer is deprived of two vinyls and one pendant vinyl becomes part of the polymer. This approach has been pursued in a number of papers.[7-11] In the binary case (si), an analytical solution for the dependence of composition on conversion is possible.[10] Also the case sd can be treated as a ternary system. The ternary copolymerisation can be quite generally described by the following scheme of propagation steps:

$$M_1^{\cdot} + M_1 \rightarrow M_1^{\cdot} \; (k_{11})$$
$$M_1^{\cdot} + M_2 \rightarrow M_2^{\cdot} \; (k_{12})$$
$$M_1^{\cdot} + M_3 \rightarrow M_3^{\cdot} \; (k_{13})$$
$$M_2^{\cdot} + M_1 \rightarrow M_2^{\cdot} \; (k_{21})$$
$$M_2^{\cdot} + M_2 \rightarrow M_2^{\cdot} \; (k_{22})$$
$$M_2^{\cdot} + M_3 \rightarrow M_3^{\cdot} \; (k_{23})$$
$$M_3^{\cdot} + M_1 \rightarrow M_1^{\cdot} \; (k_{31})$$
$$M_3^{\cdot} + M_2 \rightarrow M_2^{\cdot} \; (k_{32})$$
$$M_3^{\cdot} + M_3 \rightarrow M_3^{\cdot} \; (k_{33}) \tag{1}$$

The simplified steady-state condition,[13] according to which the rate of formation of the radical $M_n^{\cdot}$ from the radical $M_m^{\cdot}$, equals that of formation of the radical $M_m^{\cdot}$ from the radical $M_n^{\cdot}$, leads to a simple expression for the relative change in the molar concentration of vinyls M_1, M_2 and M_3, where M_1 refers to the monovinyl and M_2 and M_3 to the vinyls in the divinyl molecule, respectively. For independent reactivities (case asi in Fig. 3) one obtains:[10]

$$\frac{dM_1}{dM_2} = \frac{r_{12}M_1(M_1 + M_2/r_{12} + M_3/r_{13})}{r_{21}M_2(M_1/r_{21} + M_2 + M_3/r_{23})} \tag{2}$$

$$\frac{dM_2}{dM_3} = \frac{r_{23}M_2(M_1/r_{12} + M_2 + M_3/r_{23})}{r_{32}M_3(M_1/r_{31} + M_2/r_{32} + M_3)} \tag{3}$$

KAREL DUŠEK

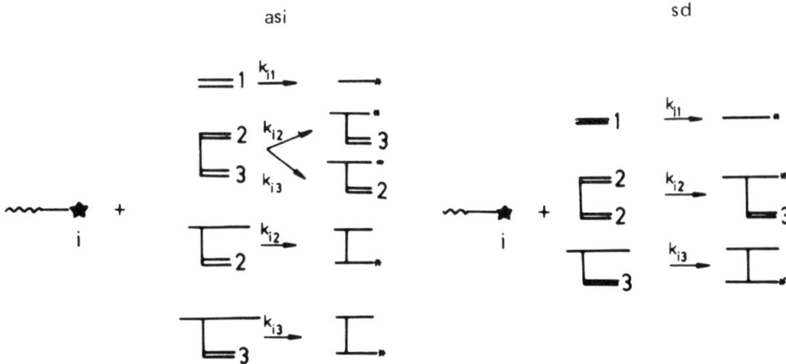

FIG. 3. Scheme of propagation steps in cross-linking copolymerisation of a monovinyl monomer with an asymmetric divinyl monomer with independent reactivities of vinyls (asi) and symmetric divinyl monomer with dependent reactivities of vinyls (sd).

where $r_{ij} = k_{ii}/k_{ij}$. Because of the independent reactivity of the vinyls, one can obtain the concentrations of singly reacted divinyl units in the copolymer m_2 and m_3 with pendant vinyls of types 3 and 2, respectively, and the concentration of doubly reacted divinyl units $m_{23} = m_\rho$ from the independent probabilities of finding reacted vinyls of types 2 and 3:

$$m_2 = (1 - \xi_3)\xi_2 M_2^0 \tag{4}$$

$$m_3 = (1 - \xi_2)\xi_3 M_2^0 \tag{5}$$

$$m = \xi_2\xi_3 M_2^0 \tag{6}$$

where M_2^0 is the initial molar concentration of vinyls of the divinyl monomer and ξ_i the conversion of the vinyl i.

The fraction of units in polymer chains, belonging to the doubly reacted divinyl units, which is usually called cross-linking density, is then given by:[10]

$$\rho = 2m_\rho/(m_1 + m_2 + m_3 + 2m_\rho) = 4\xi\Phi_2\Phi_3/\varphi_{23}^0 \tag{7}$$

where Φ_i is the integral fraction of reacted vinyls of type i, the molar fraction $\varphi_{23}^0 = M_2^0/(M_2^0 + M_1^0)$ and ξ is the molar conversion of vinyls. If the divinyl monomer is symmetric with independent reactivities of vinyl (case sd in Fig. 3), eqns. 2 and 3 degenerate to the usual binary copolymerisation equation

$$\frac{dM_1}{dM_2} = \frac{M_1(r_1M_1 + M_2)}{M_2(r_2M_2 + M_1)} = \frac{f_1(r_1f_1 + 1)}{f_1 + r_2} \tag{8}$$

where $f_1 = M_1/M_2$.

For a symmetric divinyl monomer with dependent reactivities of vinyls (Fig. 3), the kinetic scheme (1) yields

$$\frac{dM_1}{dM_2} = \frac{r_{12}M_1(M_1 + M_2/r_{12} + M_3/r_{13})}{2r_{21}M_2(M_1/r_{21} + M_2 + M_3/r_{23})} \tag{9}$$

$$\frac{dM_3}{dM_2} = \frac{r_{32}M_3(M_1/r_{31} + M_2/r_{32} + M_3)}{2r_{23}M_2(M_1/r_{21} + M_2 + M_3/r_{23})} - \frac{1}{2} \tag{10}$$

The pendant vinyl acquires the reactivity of type 3 as soon as one vinyl in the divinyl monomer has reacted. If it reacts, two vinyls of type 2 disappear from the divinyl molecule and one of these is transformed into a vinyl of type 3. Therefore, the composition of the polymer is given by the following differential equations:

$$dm_1/dm_2 = 2(dM_1/dM_2) \tag{11}$$
$$dm_3/dm_2 = -2(dM_3/dM_2) + 1 \tag{12}$$

and the cross-linking density ρ by

$$\rho = 2m_3/(m_1 + m_2 + 2m_3) \tag{13}$$

The system of differential equations 9–12 can be solved numerically to obtain the dependence of the composition and ρ on conversion. An example of expected variations in Fig. 4 shows that the cross-linking density may even pass through a maximum, if the reactivity of the vinyls in the divinyl monomer is higher than that in the monovinyl monomer.

Wesslau[9] examined the case of a low concentration of the symmetric divinyl monomer, i.e. $M_2^0 \ll M_1^0$, so that the initial concentration of the monomers $M^0 \approx M_1^0$ and of radicals $M^{\cdot} \approx M_1^{\cdot}$. An analytical solution is then possible yielding M_2, M_3, and m_3 as a function of the molar conversion of the vinyls

$$M_2 = M_2^0(1 - \xi)^{2/r_{12}} \tag{14}$$

$$M_3 = 2M_2^0/(2r_{13} - r_{12})r_{12}(1 - \xi)^{1/r_{13}} - r_{13}(1 - \xi)^{2/r_{12}} \tag{15}$$

$$m_3 = [2M_2^0/(2r_{13} - r_{12})][(r_{12}/2)(1 - \xi)^{2/r_{12}}$$
$$- r_{13}(1 - \xi)^{1/r_{13}} + r_{13} - r_{12}/2] \tag{16}$$

The composition of the polymer chain Φ_2 as well as the cross-linking density can be obtained from

$$\Phi_2 = (m_2 + 2m_3)/\xi M^0 \qquad \text{and} \qquad \rho = 2m_3/\xi M^0 \tag{17}$$

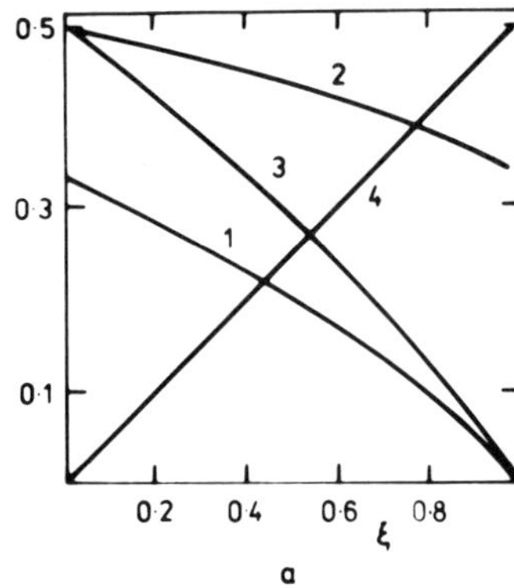 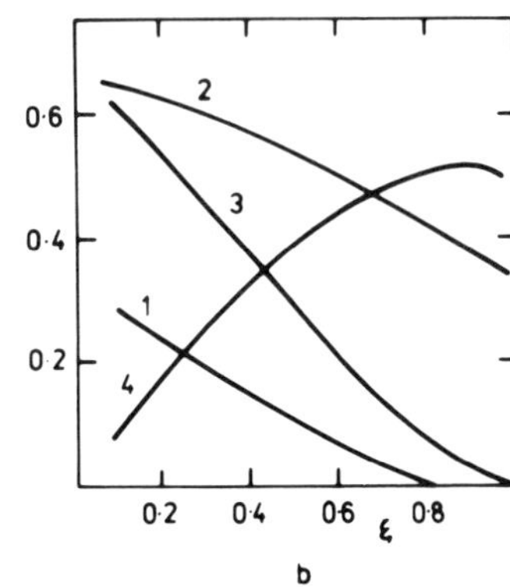

FIG. 4. Calculated composition of the monomers and polymer and cross-linking density as a function of conversion for the copolymerisation of a monovinyl monomer with a symmetric divinyl monomer with independent reactivities of vinyls. (1) Molar fraction of the divinyl monomer in the monomer mixture, (2) molar fraction of all divinyl molecules in the polymer, (3) molar fraction of units with a pendant vinyl, (4) cross-linking density ρ. (a) Copolymerisation parameters $r_1 = r_2 = 1$, initial volume fraction of monomers $\varphi_2^0 = 0.5$. (b) $r_1 = 0.5$, $r_2 = 1.0$, $\varphi_2^0 = 0.5$.

The conversion of the double bonds of the divinyl monomer, ξ_2, relative to the molar conversion of all vinyls, ξ, is given by

$$\xi_2/\xi = \Phi_2 M^0/2M_2^0 \tag{18}$$

and it is possible to show[9] that

$$\lim (\xi_2/\xi) = 1/r_{12} \tag{19}$$

which is a convenient way of obtaining the copolymerisation parameter r_{12} by extrapolation of the r.h.s. of eqn. 18 to $\xi \to 0$, and by means of eqns. 18 and 14–16 also the parameter r_{13}.

Therefore, the determination of copolymer composition as a function of conversion at low concentrations of the divinyl monomer is a convenient way of characterising the reactivity of the vinyl in the divinyl monomer as well as of the pendant vinyl.

The copolymerisation behaviour is usually determined from the polymer composition at low conversions (no compositional drift) and eqn. 8 has been widely used for determination of the reactivity ratios $r_1 = r_{12}$ and $r_2 = r_{21}$, because dM_1/dM_2 is equal to F_1^0, the initial ratio of reacted vinyls in the copolymer. At very low conversions and in the absence of cyclisation, the pendant vinyls should not enter the copolymerisation reaction, because their concentration in the system is negligibly low. This method has been widely used in the literature for determination of the copolymerisation parameters,[6-12,14-18] although it is basically incorrect due to cyclisation.

Therefore, the published copolymerisation parameters should be reconsidered. Although not discussed here, it should be noted that the treatment of binary vinyl–vinyl copolymerisation data by the methods of Finemann and Ross or Lewis and Mayo meets with some difficulty (cf. for example ref. 19).

2.1.1. Gelation

The ring-free cross-linking copolymerisation can further be developed to yield various characteristics of the reaction products, like molecular weight averages, the critical conversion at the gel-point, characteristics of the sol and gel fraction, etc. Statistical or kinetic methods can be used, but in contrast to network formation by step polyaddition or cross-linking of primary chains, the treatment of chain polymerisation is much less developed for two main reasons: (a) the existing and often conversion-dependent degree of polymerisation and compositional distribution of primary chains, and (b) complications due to a strong cyclisation making the application of the perturbation method to the ring-free case of little use. Moreover, the network formation by cross-linking chain reactions is a kinetically controlled process and application of statistical methods is only an approximation to the real situation[20] (it may, however, be good). The application of the kinetic method was demonstrated by Kuchanov and Pismen.[21]

The statistical methods are all based on the tree-like model and the trees are either generated by linking monomer units[22] or by linking primary chains via a random combination of cross-linked units.[2,6,23] The former method has been developed as yet only for a random reaction (equal and independent reactivities): it is more difficult and for the unequal reactivity case requires the use of Markovian statistics and a correct copolymerisation model. Data on the initial composition and copolymerisation kinetic constants are the input parameters. The method based on primary chains is less rigorous, but relatively simple. Here, the degree of polymerisation and compositional distributions of primary chains (obtained either experimentally or by Markov chains methods[24]) and the fraction of cross-linked units obtained from copolymerisation kinetics are the input parameters.

The gel-point condition is based on the requirement that the branching probability is unity, which means in the language of the theory of branching processes that the value of the first derivative of link probability generating function for units in the first and higher generations is unity. This condition yields for the gel-point

$$\frac{\sum\limits_{P,\rho} P^2\rho^2 n_{P,\rho}}{\sum\limits_{P,\rho} P\rho n_{P,\rho}} - \frac{\sum\limits_{P,\rho} P\rho^2 n_{P,\rho}}{\sum\limits_{P,\rho} P\rho n_{P,\rho}} = 1 \qquad (20)$$

where $n_{P,\rho}$ is the number fraction of primary chains composed of P monomer units of which the fraction ρ is cross-linked and P refers to the primary distribution (later called P^0) with averages P_n^0, P_w^0, etc. The primary chains, with respect to their degree of polymerisation and composition, are chains obtained by severage of all cross-links, e.g. by chemical splitting. They are sometimes approximated by chains obtained from the corresponding monovinyl monomer prepared under identical conditions— an approximation tolerable at low concentrations of the divinyl monomer. If $P_n^0 \gg 1$, the second term in eqn. 20 becomes negligible and the first term is the 'weight' average $\langle P^0\rho \rangle_w$, so that $\langle P^0\rho \rangle_w = 1$.

If the degree of polymerisation and compositional distributions are mutually independent, which may be approximately true for long primary chains, the gel-point conditions become

$$\rho_w(P_w^0 - 1) = 1 \qquad \rho_w = \sum\limits_{P,\rho} \rho^2 n_{P,\rho} \Big/ \sum\limits_{P,\rho} \rho n_{P,\rho} \qquad (21)$$

and if moreover ρ is equal and constant for all primary chains (long chains, low conversion)

$$\rho(P_w^0 - 1) = 1 \qquad (22)$$

The gel-point equation is usually employed in this form, but if gelation occurs at higher conversion or if the compositional distribution is wide for other reasons, $P_w^0\rho \neq P_w^0\rho_w \neq \langle P^0\rho \rangle_w$.

Employing eqn. 22, the critical conversion is obtained using a relation between ρ and ξ (e.g. eqn. 7). For the symmetric independent case $\rho = \xi\Phi_2^2/\varphi_2^0$; for the random case $\Phi_2 = \varphi_2^0$ and then the critical conversion of vinyls ξ_c is given by

$$\xi_c = [\varphi_2^0(P_w^0 - 1)]^{-1} \qquad (23)$$

An extension of the treatment to include monomers with more than two double bonds is easy.[22] Once formulated, the probability generating functions employed for the gelation behaviour can be exploited for the calculation of other pre-gel and post-gel structural parameters of the system in a routine way.

2.2. Copolymerisation with Cyclisation

As explained in section 1, cyclisation is expected to play an important role in chain cross-linking copolymerisation and this expectation has been verified experimentally in a number of papers. Both the low and high conversion behaviour is affected.

In the treatment of cyclisation, one sometimes talks about cyclisation when the growing radical attacks the pendant double bonds in the same primary chain and about what Holt and Simpson[25] call multiple cross-linking, if the radical attacks double bonds pendant on other chains already chemically connected with the growing macroradical (Fig. 1). Cyclisation involving double bonds pendant on the same growing macroradical is indeed the only source of rings at the start of the reaction, but later on the distinction between double bonds of the same molecule and the chemically attached chains is meaningless and so cyclisation is understood to be every chemical connection of two pendant double bonds already connected by at least a single path of chemical bonds. It is clear that the flexible growing macroradical can attack pendant double bonds at different distances along the connecting paths and so rings of different sizes can be formed.

The introduction of cyclisation steps into the copolymerisation scheme can be simplified by introducing a single cyclisation constant k_c as Aso[26] did for the polymerisation of a divinyl monomer.

$$
\begin{array}{ccc}
\begin{array}{c} -B^{\cdot} \\ | \\ B \end{array} & + & \begin{array}{c} B \\ | \\ B \end{array} \qquad & M_2^{\cdot} + M_2 \rightarrow M_2^{\cdot}\,(k_{22})
\end{array}
$$

$$
\begin{array}{ccc}
\begin{array}{c} -B^{\cdot} \\ | \\ B \end{array} & + & \begin{array}{c} -B- \\ | \\ B \end{array} \qquad & M_2^{\cdot} + M_3 \rightarrow M_3^{\cdot}\,(k_{23})
\end{array}
$$

$$
\begin{array}{ccc}
\begin{array}{c} -B- \\ | \\ B^{\cdot} \end{array} & + & \begin{array}{c} B \\ | \\ B \end{array} \qquad & M_3^{\cdot} + M_2 \rightarrow M_2^{\cdot}\,(k_{32})
\end{array}
$$

$$
\begin{array}{ccc}
\begin{array}{c} -B- \\ | \\ B^{\cdot} \end{array} & + & \begin{array}{c} -B- \\ | \\ B \end{array} \qquad & M_3^{\cdot} + M_3 \rightarrow M_3^{\cdot}\,(k_{33})
\end{array}
$$

$$
\begin{array}{cc}
\begin{array}{c} -B^{\cdot} \\ | \\ B \end{array} \qquad & M_2^{\cdot} + M_3 \rightarrow M_3^{\cdot}\,(k_c) \qquad (24)
\end{array}
$$

The differential equations corresponding to this scheme can be solved analytically and by a proper choice of experimental conditions the extent of cyclisation can be found. However, such a modification is of little value unless the last step is the only source of cyclisation, i.e. unless cycles are closed only within monomer units and not at a longer distance. Such a situation may arise in sterically preferred cyclopolymerisation. In the other cases, however, the $M_2^.$ radical also comes into play by attacking neighbouring and more distant double bonds. The other difficulty arises from the fact that cyclisation 'constants' are dependent on conversion, a fact following from the conversion-dependent change in the concentration of available reaction partners—pendant double bonds.

A similar approach to copolymerisation has been used by Roovers and Smets,[27] or Matsumoto et al.[28] They also considered ring closure within the bis-unsaturated units to be the only source of cyclisation, but they refined the scheme by ascribing different reactivities to the cyclised and non-cyclised macroradicals. Indeed, small rings can exert a steric influence on the reactivity of the radical, but the effect becomes weaker when the rings get larger.

Instead of speaking of a single or several cyclisation constants for individual steps in the reaction scheme, one should rather examine the probability of the growing macroradical approaching a double bond already connected with this macroradical.

It was shown and experimentally proved by many authors that ring closure in linear and weakly branched systems is controlled by the conformational statistics of the sequence of bonds connecting the reacting functionalities.[29] The application of conformational statistics based on the rotational-isomeric state model was relatively successful (cf. for example ref. 30). Except for the case when small rings predominate, the Gaussian distribution function approximation of the equivalent chain for the end-to-end distance works satisfactorily.

The situation in polyfunctional systems is more complicated because of steric effects of the existing rings on ring closure (correlations due to cycles). The kinetic and statistical approaches disregard correlations due to cycles and the tree-like model remains essentially untouched although properly perturbed to take into account bonds (cross-links) that are not effective in branching. Spatial correlations including those due to cycles can be adequately expressed by percolation models,[31–33] but at present the models are not yet sufficiently developed to take properly into account the chain flexibility and specificity of the cross-linking reaction. Despite the neglect of correlation due to cycles and possible excluded volume effects, the approximation of ring-closure probability by the conformational statistics of a single connecting path of the perturbed tree-like model proved useful also in polyfunctional systems in the case of mild cyclisation (cf. for example accounts by Gordon *et al.*[34-37] or Stepto[38]).

2.2.1. Cyclisation Probability

The probability of a contact between two functionalities can be described by the Gaussian distribution function with good approximation if they are separated by more than about 20 links, but this limit depends on the chain flexibility (for stiff chains the limit may be much higher).

In the Gaussian approximation describing the conformational statistics of the equivalent chain, the probability $W(R)dV$ that one end of a chain, composed of n_s statistical segments of length l_s each, resides in the volume element dV at a distance R ($R^2 = X^2 + Y^2 + Z^2$), if the other end is fixed at the origin of the coordinate axes, is given by[2]

$$W(R)dV = (3/2\pi n_s l_s^2)^{3/2} \exp(-3R^2/2n_s l_s^2)dV \tag{25}$$

The distribution function 25 is valid for any two segments of a chain separated by n_s segments. The formation of a bond is possible, if the chain segments bearing unreacted functionalities meet in the volume element $dV(R=0)$. Selecting an unreacted functionality at random, the local concentration of unreacted functionalities supplied by a segment (monomer unit) separated by x monomer units means at the same time the probability P_{cx} of closure of a ring composed of x monomer units

$$P_{cx} = W_x(0)m_x/N_A = \Omega[(3/2\pi Q l^2)^{3/2}/N_A](xb)^{-3/2}m_x \tag{26}$$

where Ω is the average number of unreacted functional groups per monomer unit, attached to the unit with randomly selected functionality, Q is the number of main chain atoms in a statistical segment, l is the length of the main chain bond, N_A is the Avogadro number and b is the number of atoms which a monomer unit contributes to the main chain; m_x is the average number of monomer units at a distance of x units from the unit with preselected functionality (Fig. 5).

The probability P_{cx} can be rewritten as

$$P_{cx} = \Omega\Lambda(xb)^{-3/2}m_x \tag{27}$$

where

$$\Lambda = (3/2\pi Q l^2)^{3/2}/N_A$$

characterises the chain flexibility and therefore the tendency to cyclisation. The $-3/2$ law shows the decreasing tendency to cyclisation when the separation gets larger.

KAREL DUŠEK

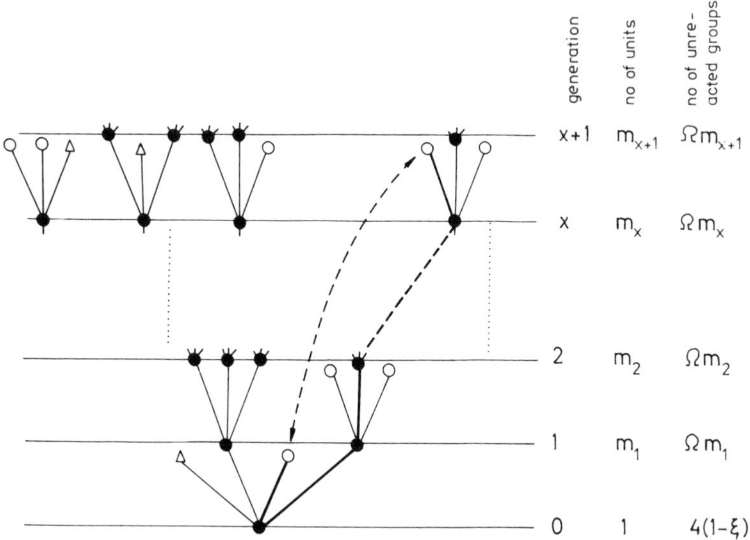

FIG. 5. Sketch of the spanning-tree approach to cyclisation. ● Nodes representing monomer units, ○ terminal nodes representing unreacted functional groups, △ reacted functional groups engaged in intra-molecular bonds. Reaction of two ○ yields two △.

Considering the randomly selected functional group, its reaction partners can be supplied by monomer units at various separation distances x, so that the total cyclisation probability is given by the sum over all possible x

$$P_c = \Omega\Lambda \sum_{x=x_t}^{\infty} (xb)^{-3/2} m_x \qquad (28)$$

To refine the model, one can calculate separately the ring closure probabilities for the smallest ring(s) and perform summation starting from the threshold value of xb sufficiently large for use of the Gaussian approximation. The parameter Λ can be obtained, for example, from solution studies, and m_x from the statistical treatment of the cross-linking reaction (cf. for example ref. 1), since it is equal to the number of monomer units in the generation x.

A similar statistical approach to cyclisation in the chain cross-linking polymerisation of diallyl phthalates was used for the first time by Haward,[39] and later by Holt and Simpson,[25] to calculate the proportions of cyclised units. The agreement with experiments was of the right order of magnitude. However, the possibility of cyclisation along the single path of the primary chain was considered, an approximation that is good for the initial stages of reaction. The statistical treatment of cyclisation in the chain polymerisation of a bis-unsaturated monomer was later extended to take into account the dependence of the extent of cyclisation on polymerisation conversion, length of primary chains, dilution, etc.[40–41] The treatment was based on the theory of branching processes and spanning-tree approximation. As can be seen from eqn. 28, the cyclisation probability depends on $m_x x^{-3/2}$, which absorbs contributions to cyclisation coming from ring closure on the same primary chain (and is therefore dependent on its degree of polymerisation), as well as from all attached chains in successive generations (multiple cross-linking) which become increasingly important when the polymerisation conversion and therefore the degree of branching increase.

Model calculations[40] show that the fraction of cyclised units in the polymer σ increases steeply up to a degree of polymerisation of primary chains of about 20 and reaches a saturation value at $P \approx 75$–100. The inclusion of multiple cross-linking makes a contribution to σ of about 15 % at $P \approx 50$ and less at lower P. The agreement with experiment (residual unsaturation, critical conversion) was reasonable only if one assumed an unrealistically high flexibility ($Q = 2$) of the chains. The general applicability of this model to polymerisation of bis-unsaturated monomers will be discussed in the following section.

2.2.2. Cyclisation and Determination of Copolymerisation Parameters

In the determination of copolymerisation parameters, the situation is better in that the experiments can be limited to low conversions, where the compositional drift and cross-linking effects can be neglected. However, cyclisation still plays an important role and must affect the determination of the copolymerisation parameters. In spite of this, the effect has been ignored and (except for some work on cyclocopolymerisation) the overwhelming majority of published reactivity ratios were obtained assuming the ring-free reaction.

The cyclisation reaction in low conversion copolymerisation has been taken into account by considering ring closing reactions with participation of all kinds of free radicals but assuming the same reactivity of the cyclised and non-cyclised radical of the divinyl component[42] (this restriction could be relaxed if necessary). Copolymerisation of a symmetric divinyl monomer with independent reactivities of vinyls has been considered, but again the condition of independence could be relaxed if necessary. If the molar fractions of reacted vinyls are denoted by Φ_1 (in monovinyl monomer), Φ_2 (in divinyl units with pendant vinyls) and Φ_3 (in divinyl units with both reacted vinyls), where $(\Phi_1 + \Phi_2 + \Phi_3 = 1)$, the rate equations are given by

$$v_1 = dC_p\Phi_1/dt = \varphi_1 C_0(k_{11}M_1^{\cdot} + k_{21}M_2^{\cdot} + k_{31}M_3^{\cdot}) \tag{29}$$

$$\begin{aligned} v_2 = dC_p\Phi_2/dt = &-\varphi_3 C_0(k_{13}M_1^{\cdot} + k_{23}M_2^{\cdot} + k_{33}M_3^{\cdot}) \\ &- P_c\Phi_2(k_{13}M_1^{\cdot} + k_{23}M_2^{\cdot} + k_{33}M_3^{\cdot}) \\ &+ \varphi_2 C_0(k_{12}M_1^{\cdot} + k_{22}M_2^{\cdot} + k_{32}M_3^{\cdot}) \end{aligned} \tag{30}$$

$$\begin{aligned} v_3 = dC_p\Phi_3/dt = &\, 2C_0\varphi_3(k_{13}M_1^{\cdot} + k_{23}M_2^{\cdot} + k_{33}M_3^{\cdot}) \\ &+ 2P_c\Phi_2(k_{13}M_1^{\cdot} + k_{23}M_2^{\cdot} + k_{33}M_3^{\cdot}) \end{aligned} \tag{31}$$

where v_1, v_2 and v_3 are velocities, C_0 and C_p are the initial number of vinyls and number of reacted vinyls, respectively, and $P_c\Phi_2$ is the cyclisation probability equivalent to P_c in eqn. 28. Thus, units with pendant vinyls (Φ_2) disappear either due to cross-linking reaction with the rate proportional to $C_0\varphi_3$ or cyclisation reaction (proportional to $P_c\Phi_2$); both reactions produce reacted vinyls in doubly reacted divinyl units (Φ_3).

At low conversions, the molar fraction of pendant vinyls, φ_3, is negligibly small with respect to φ_1 and φ_2, and from the condition of symmetry and independence it follows that $k_{i3} = k_{i2}$ and $k_{3j} = k_{2j}$, so that the final copolymerisation equation:[42]

$$F_1 = f_1 \frac{f_1 r_1 + 1 + Y}{f_1 + r_2 + Y r_2} \tag{32}$$

resembles that for ring-free binary copolymerisation, eqn. 8, which has been widely used for determination of copolymerisation parameters in vinyl–divinyl copolymerisation. In eqn. 32, the extra terms Y and $Y r_2$ take account of cyclisation:

$$Y = P_c\Phi_2/\varphi_2 C_0 \tag{33}$$

where P_c is a cyclisation constant averaged over all possible lengths of the growing macroradical.[42] Since eqn. 32 is valid at low conversions, contribution to cyclisation from primary chains attached to the growing macroradical need not be considered.

The parameter Y can be obtained directly from experimentally measured Φ_2 and φ_2 or from the composition of the polymer and monomer F_1 and f_1 employing the relationship

$$\Phi_2 = \{[1 + (R_2/f_1)(1 + Y)](1 - Y)\}/\{f_1 r_1 + (1 + Y)[2 + (r_2/f_1)(1 + Y)]\} \tag{34}$$

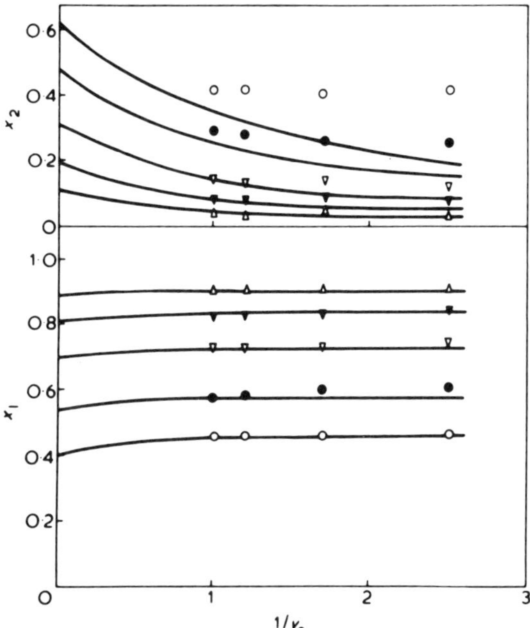

FIG. 6. Dependence of molar fraction of styrene units (x_1) and ethylene dimethacrylate units with pendant double bond (x_2) in low conversion ST–EDMA copolymers on dilution;[42] v_0 is the volume fraction of the monomers in the initial mixture. Initial weight fraction of EDMA: △, 0·05; ▼, 0·10; ▽, 0·20; ●, 0·40; ○, 0·60. The curves were calculated according to the model in section 2.2.2 (details in ref. 42).

By calculating the parameter P_c using, for example, Gaussian conformational statistics, one can get reactivity ratios unperturbed by cyclisation, but the best way is to make use of the fact that Y varies with dilution. Increasing dilution makes C_0 smaller and Y higher, and so by extrapolating to the experimentally inaccessible region $1/C_0 \to 0$, one can reduce Y to zero. The dependence of F_1 can be expanded into a power series of $1/C_0$

$$F_1 = a_0 + a_1/C_0 + a_2/C_0^2 + \cdots \tag{35}$$

where a_0 is equal to the r.h.s. of eqn. 32 with $Y = 0$ and a_1 is a function of P_c.† Unfortunately, the plot of F_1 versus $1/C_0$ is not linear (Fig. 6) and so the determination of the extrapolated value is somewhat difficult. However,

† In ref. 42, there is a regrettable misprint: b_1 in eqn. A12 should read $b_1 = [(f_1 + r_2)/(f_1^2 r_1 + 2f_1 + r_2)](P_c/\varphi_2)$.

164   KAREL DUŠEK

regression analysis may help in determining both extrapolated F_1 and P. The deviations coming from the neglect of cyclisation using eqn. 32 in the determination of copolymerisation parameters have not yet been analysed for a wider range of reactivity ratios, but only in the case of the copolymerisation of styrene with ethylene dimethacrylate—a system that does not deviate much from ideality (for styrene–methyl methacrylate, $r_1 = 0.51–0.55$, $r_2 = 0.49–0.58$). Figure 6 shows that the expected composition given by F_1 does not change much with dilution with an inert diluent, but that the expected decrease in the fraction of units with pendant double bonds is appreciable.

Beyond the limit of zero conversion, the situation becomes more difficult due to a dependence of the cyclisation intensity on conversion and due to the compositional drift. A statistical treatment is possible and can be based either on the concept of primary chains[40,41] or monomer units[22] as roots of the probability trees.

2.3. Copolymerisation Models: Conclusions

At present, there exists an exact ring-free kinetic description of the vinyl–divinyl copolymerisation, but the results are of little use due to the neglected cyclisation. However, the results can be useful for comparative estimation of the cyclisation effect and they can also, at least semiquantitatively, predict the—sometimes decisive—effect of copolymerisation parameters on compositional heterogeneity and network formation. Until now, the majority of copolymerisation parameters, except those on cyclopoly-merisation, have been obtained by neglecting cyclisation and they may not, therefore, reflect correctly the real reactivity ratios determined by the chemical structure of the monomers. The inclusion of long-range cyclisation steps into copolymerisation and network formation has been initiated and can be further developed.

It should be remembered, however, that the present theories are in fact mean field theories based on the equal chance of every double bond to be engaged in the reaction with a free radical. An analysis of experimental data which follows will show that this assumption is fulfilled only in some cases.

3.   EXPERIMENTAL STUDIES OF (CO)POLYMERISATION

The theoretical models (usually ring-free) of cross-linking copolymerisation kinetics and statistics offer various structural characteristics which can be compared with experiments. These characteristics are, for example, the composition of the (co)polymer, the fraction of doubly reacted cross-linker units (cross-links), the gel fraction after passing the gel-point and the fraction of effective cross-links.

3.1. Pendant Double Bonds

All ring-free models predict that every bis-unsaturated unit incorporated in the polymer should initially bear a pendant double bond. As the polymerisation proceeds, the pendant double bonds are gradually converted into cross-links, unless they are prevented from doing so at some stage of the reaction due to steric reasons or overall diffusion control connected with the glass transition.

Experimental results give clear evidence that this prediction is not fulfilled: only a part of the bis-unsaturated units in the (co)polymer bear pendant double bonds and the fraction of doubly reacted units is considerable. The decrease in the fraction of units having pendant double bonds with conversion is much less rapid than predicted by the ring-free models, and eventually a considerable number of units with pendant double bonds survive in the (co)polymer and do not polymerise any more.

Diallyl compounds, especially diallyl phthalates and some divinyl compounds (ethylene dimethacrylate, EDMA, and divinylbenzene, DVB), were among the most studied compounds in *homopolymerisation*. It is well known that allyl monomers undergo a degradative chain transfer and the resulting degree of polymerisation is low, amounting to several tens. Therefore, the gel-point conversion is relatively high, which makes possible a study of the soluble polymers in a broader conversion range. Simpson and Holt[25,43] found that in diallyl orthophthalate polymers (DAP) only about one-third of the units bear pendant double bonds. At the degree of polymerisation of primary chains $P_n^0 = 20$, the unsaturation x changes from 0.26 to 0.29 when the weight conversion goes up from 3 to 23%. The unsaturation, x, is related to the unsaturation of the DAP monomer, so that $x = 1/2$ corresponds to the situation when all units bear pendant double bonds. One can express the amount of pendant allyl groups by the fraction p. The primary chain is composed of the fraction p of units bearing a pendant double bond, s of units engaged in intra-molecular cross-links and a of units engaged in inter-molecular cross-links ($p + s + a = 1$). Then

$$p = x/(1 - x) \qquad (36)$$

so that $p = 0.35$–0.41.

Alternative measurements gave a somewhat higher unsaturation x ranging from 0.45 to 0.34 for $P_n^0 = 28$, but the most recent detailed measurements of Matsumoto *et al.*[45] (cf. Table 1 and Fig. 7) are in agreement with the original data of Simpson and Holt. The combined data show that:

(a) unsaturation decreases only slowly with increasing polymerisation conversion; the pendant vinyls are partly consumed in intermolecular cross-linking and partly in cyclisation increasing due to multiple cross-linking;

(b) unsaturation decreases with increasing dilution because of the discussed promotion of cyclisation by dilution (Fig. 7). With increasing dilution, the concentration dependence of x gets weaker. The values of x decrease linearly with increasing concentration of the monomer;

(c) not much can be said about the effect of molecular weight of primary chains. The unsaturation should mildly decrease with increasing molecular weight of primary chains due to an increased opportunity for cyclisation, but data in a sufficiently broad range of P_n^0 are not available;

(d) x is markedly affected by the length and stiffness of the monomer unit determining the probability of ring closure.[25,35] Among the series of diallyl esters, diallyl orthophthalate[25] (DAP) yields the lowest value of x, whereas for diallyl terephthalate,[25] x is as high as 0·41–0·45 so that only 2–3 out of 10 units are engaged in cycles. This fact follows from the steric impossibility of forming a ring within a single DAT unit and a lower P_c due to increased stiffness of the bridge.

Unlike the phthalate polymers, the vinyl polymers have usually a much higher degree of polymerisation, unless a larger amount of chain-transfer agent is used. Because of this fact, the gel-point is shifted to lower conversions and the major part of the network build-up occurs after the gel-point has been passed.

In contrast to a much higher P_n^0, which should promote cyclisation and lower the concentration of pendant vinyls, unsaturation is much higher than in diallyl phthalate polymers. So, for low conversion soluble ethylene dimethacrylate (EDMA) polymers,[46] p reaches the value 0·7 (cf. also ref. 26), and in p-divinylbenzene (p-DVB) polymers,[47] p is about 0·5 at conversions up to 20 % by wt.

From dilatometric measurements and i.r. analysis, however, Loshaek and Fox[48] found the cross-linking efficiency to be 30 % in high conversion EDMA polymers, which corresponds to $p = 0.7$ as in low conversion polymers.

TABLE 1

UNSATURATION, x, GEL-POINT CONVERSIONS AND CROSS-LINKING EFFICIENCY, α, IN THE BULK POLYMERISATION OF DIALLYL ORTHOPHTHALATE (DAP), DIALLYL ISO-PHTHALATE (DAI) AND DIALLYL TEREPHTHALATE (DAT)[45]

ξ_w	x	p	P_n	P_n^0	α
DAP					
0·042	0·278	0·385	32·8	28·1	0·017
0·093	0·271	0·370	34·3	27·9	0·028
0·126	0·266	0·362	35·8	26·5	0·032
0·172	0·260	0·351	41·6	25·2	0·048
0·228	0·255	0·342	48·3	22·7	0·073
DAI					
0·052	0·402	0·670	32·0	32·5	—
0·084	0·394	0·650	35·7	34·7	0·009
0·116	0·388	0·634	39·9	33·6	0·028
0·160	0·381	0·615	45·4	30·1	0·056
0·202	0·373	0·595	63·5	32·3	0·080
DAT					
0·057	0·450	0·818	32·9	35·6	—
0·120	0·442	0·792	38·7	34·7	0·014
0·184	0·435	0·770	49·2	33·6	0·041
0·233	0·429	0·751	64·8	33·1	0·061

GEL POINTS

Monomer	ξ_w	p	P_w^0	α
DAP	0·223	0·34	64	0·094
DAI	0·218	0·59	62	0·120
DAT	0·218	0·75	64	0·125

ξ_w is the weight conversion of the monomer, x is unsaturation of the polymer related to unsaturation of the monomer, p is the fraction of units of primary chains bearing pendant double bonds, P_n is the number-average degree of polymerisation, P_n^0 and P_w^0, respectively, are the number- and weight-average degrees of polymerisation of primary chains (obtained by saponification), and $\alpha = a/(a + s)$ is the fraction of cross-linked units engaged in inter-molecular bonds.

The fraction of units with pendant double bonds decreases with increasing conversion: for EDMA polymers[26] p falls from 0·7 to about 0·5 at 90 % conversion. Like the case with diallyl ester polymers, dilution causes a decrease in p but a milder one: Galina *et al.*[46] reported a decrease in p from 0·7 in bulk down to 0·6 in a 30 % solution. Perhaps a somewhat steeper decrease (the conversion was variable and gelled systems were examined in

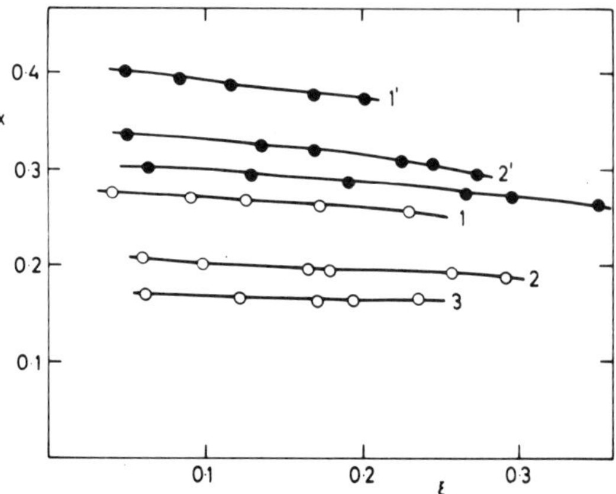

FIG. 7. Dependence of unsaturation, x, on weight conversion in the polymerisation of diallyl orthophthalate (DAP = curves 1, 2, 3) and of isophthalate (DAI = curves 1', 2').[45] Concentration of the monomer: (1),(1') 100 %, (2),(2') 50 %, (3) 25 %.

part) was found by Aso[26]—in 1 % benzene solution p had fallen to 0·1, but the primary chains must have been rather short at such a high dilution and the pendant vinyls may also have disappeared by other than polymerisation reactions.

In fact, the so-called cyclopolymerisation of bis-unsaturated monomers does not differ in principle from the polymerisation of diallyl esters, EDMA or DVB discussed above. As a rule, pendant double bonds are always present in the polymer and some of them may become involved in the intermolecular reaction. For example, p values up to 0·43 were found in polymers of o-DVB[49] in about 20 % solution in toluene or 0·24–0·28 for polymers of divinyl ether[50] in a 20 % solution in benzene. The somewhat higher tendency to cyclisation (given by the decisive contribution of small rings) of these rather than of other monomers makes the preparation of soluble polymers at a reasonable conversion (20–30 %) possible.

Experiments on *vinyl–divinyl copolymerisation* are of great importance for elucidation of the process and structure of the product, since the incorporation of the monovinyl monomer makes the structure of the polymer looser. By lowering the concentration of the divinyl monomer, the chains approach the linear form and the state of normal coiling. In order to test the theories of cyclisation, one should preferably extrapolate the experimental results to zero concentration of the divinyl monomer. As has been shown, extrapolation to zero conversion does not help, since the macromolecules may be heavily internally cross-linked. However, a sufficiently precise determination of the amount of singly and doubly reacted cross-linker molecules may become the stumbling block, due to their low concentration. Also, the possible unequal reactivities of the comonomers may become a complicating factor in that the determination of the copolymerisation parameters is complicated by cyclisation (cf. Chapter 1).

Before examining the experimental data on copolymerisation, we should repeat the predictions from the simple models. The ring-free model predicts that 100 % divinyl units bear a pendant double bond at zero conversion, irrespective of the concentration of the divinyl monomer, c_{DV}. The fraction p should decrease to zero when the reaction approaches completion. The model with conformationally determined cyclisation predicts that the fraction p is always lower than unity even at zero concentration of the divinyl monomer. However, this fraction should decrease both with increasing conversion and increasing c_{DV}. The predictions that $p < 1$ at $\xi \to 0$ and $c_{DV} \to 0$, and therefore that some cyclisation must occur in this limit, may seem somewhat strange. The explanation is to be sought in the fact that cyclisation does not occur between two pendant vinyls, but that there is a finite chance for the free radical residing on the newly added divinyl unit or any successively added monovinyl unit to coil back and attack the pendant vinyl (cf. eqn. 1).

Checking of the *composition* of the copolymer against the model is hampered by the difficulty of getting reliable chemical reactivity ratios not obscured by cyclisation or other effects (shielding). The copolymerisation parameters have been usually obtained without taking into account cyclisation[15,16] and when taken as such then of course good agreement with the ring-free model must be obtained at low conversions. Comparing the results in the whole conversion range for styrene (ST)–*m*-DVB or *p*–DVB copolymers, (8–50 % DVB), Malinský *et al.*[51] found slight negative deviations in the fraction of incorporated DVB in the polymer when its composition was determined by i.r. spectrometry, and almost full agreement when the unreacted DVB was measured by gas chromatography. These findings suggest that even a strong cyclisation (proved independently of unsaturation) may not have a strong effect on the fraction of incorporated divinyl monomer and copolymerisation parameters. The effect of cyclisation on composition was later analysed by Dušek and Spěváček[42] using the conformation model and applying it to the system ST–EDMA. Again, the effect of cyclisation even promoted by progressive dilution did not have a strong effect on composition (Fig. 6); it was estimated that the

170   KAREL DUŠEK

copolymerisation parameters for ST–EDMA may be distorted by the
neglect of cyclisation by not more than about 20 %.

The copolymerisation equation (32) predicts that the effect of
cyclisation on the determination of the copolymerisation parameters
should disappear in the limit of zero concentration of the cross-linker;
indeed

$$\lim_{\substack{f_2 \to 0 \\ \xi \to 0}} (F_1/f_1) = 1/r_1 \qquad\qquad (37)$$

which is a convenient way of obtaining the copolymerisation parameter, r_1,
which has already been applied several times.[9,52,53] However, Y may not
always be neglected with respect to f_1; in this case, r_1 calculated from the
simple relation[8] should exhibit a dependence on the fraction of the divinyl
monomer φ_2^0. Indeed, such a dependence is observed experimentally.[52,53]
The change of r_1 with φ_2^0 has the right trend, but a quantitative agreement
would require a higher value of P_c/C_0 than that obtained from
conformational statistics of the respective chains.[53] An explanation can be
given that cycles already formed (they are more numerous in copolymers
with a higher concentration of the divinyl monomer) further promote
closure of new rings, so that P_c/C_0 values calculated from chain statistics
can rather be considered the lowest bounds.

The content of *pendant vinyls* (unsaturation) is much more sensitive
towards the departures from the ring-free models. The data obtained for a
broad range of cross-linker concentrations and polymerisation conversion
show the following features:

(a) There is only a moderate dependence of the fraction of doubly reacted
cross-linker units $(a + s)$ or fraction of units with pendant vinyls (p) on
conversion and with increasing cross-linker content in the feed $(a + s)$
becomes more and more independent of conversion (examples of ST–DVB
copolymers are given in Fig. 8). Similar results were also obtained for ST
copolymerised with other cross-linkers;[52,54]

(b) The dependences of the fraction of cross-linker units doubly reacted or
with pendant double bonds on x_{DV}^0 in low conversion copolymers are
somewhat conflicting. What is sure is that the dependence of p, or $a + s =
1 - p$, approaches a steady value at higher *concentrations of the cross-
linkers*, with p varying between 0·5 and 0·8—a value which is quite high.
Discrepancies can be noticed at lower concentrations of the cross-linker
(Fig. 9). The n.m.r. analysis of ST–EDMA copolymers[42] reveals that the
fraction $1 - p$ decreases with increasing concentration of EDMA, x_{DV}^0,

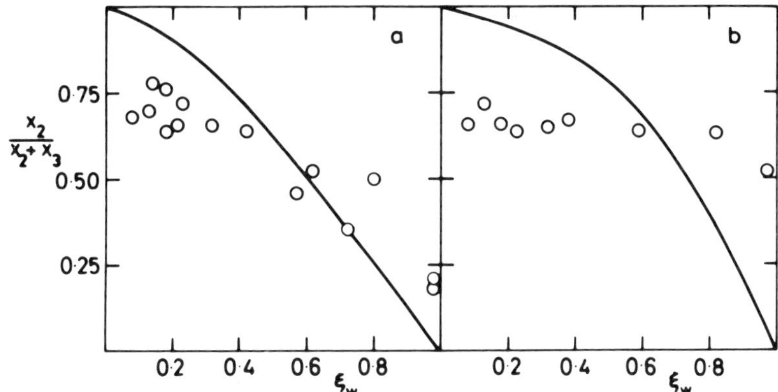

FIG. 8. Fraction of divinyl units with pendant vinyl in styrene–p-divinylbenzene copolymers as a function of weight conversion. (a) 8 wt % DVB, (b) 15 wt % DVB, ——— ring-free theory (data partly from ref. 51).

when the content of EDMA in the feed is higher than 5 wt % (cf. also Fig. 6). On the contrary, this fraction decreases with decreasing x_{DV}^0 for ST–EDMA copolymerised in 15 % solution[52,54] and may seem to extrapolate to zero at $x_{DV}^0 = 0$. The results on ST–DVB systems show a mild decrease[51] and on ST–DVB–cyclohexane systems a constant value of $1 - p$.[55] Whitney and Burchard[22] report for $1 - p$ in MMA–EDMA and MMA–PETMA copolymers a value of 0·1–0·2 and Eschwey and Burchard[56] a value of 0·6 and 0·9 for anionicly copolymerised ST + p-DVB and ST + m-DVB, respectively. Minnema and Staverman[8] report $(1 - p) \approx 0.25$ to be independent of the cross-linker concentration for copolymers with a stiff and long bridge between the methacrylic groups in the cross-linker. Also, Matsumoto *et al.*[28] claim a constant but high value of $(1 - p) \approx 0.8$ for low conversion solution copolymers of diallyl orthophthalate with allyl benzoate and vinyl acetate.

It is difficult to explain the relatively large differences, but they may be partly accounted for by inaccuracy of the analytical methods of determination of the respective fractions of divinyl units in the copolymer, if their total amount is low. The polymer separation procedure may be another source of errors: the polymer solutions and particularly the precipitated monomer should be stabilised against further polymerisation of pendant vinyls and the separated polymer must be reprecipitated at least twice, in order to remove all unreacted monomer, which cannot be removed by drying *in vacuo* at room temperature and which would then increase the apparent concentration of pendant vinyls.

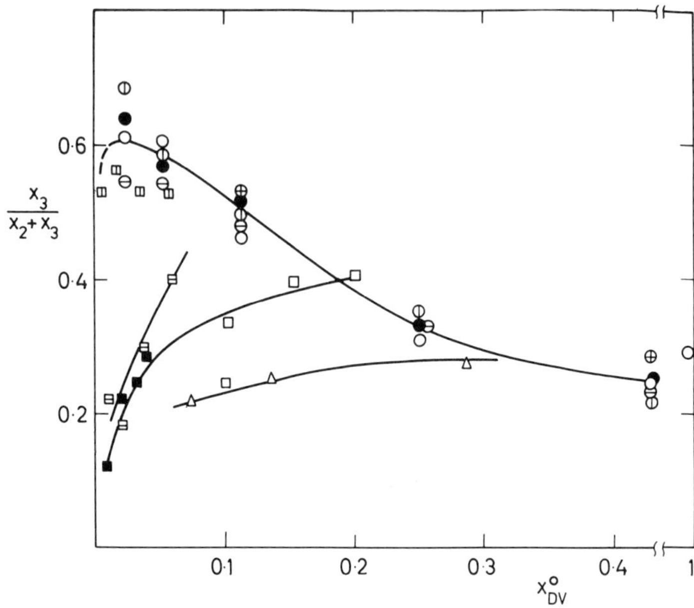

FIG. 9. Fraction of doubly reacted divinyl molecules relative to all divinyl molecules in low conversion styrene (ST)–ethylene dimethacrylate (EDMA) or ST–divinylbenzene (DVB) copolymers, as a function of molar fraction of the divinyl monomer in the feed. ST–EDMA[42] initial volume fraction of monomers; v_0: ○ 1·0, ● 0·8, ⊖ 0·6, ① 0·4, ⊕ 0·2; ST–DVB,[55] 15% solution ⊟ in toluene, ⊞ in cyclohexane; ST-p-DVB,[51] △; ST–EDMA, 15% solution in toluene, ref. 54 ■ and ref. 52 □.

(c) The relatively weak dependence (except in the region of very low x^0_{DV}) or rather independence of the fraction $1 - p$ of *dilution* at higher values of x^0_{DV} is another special feature of the cross-linking copolymerisation. Progressive dilution should favour cyclisation and make the fraction of doubly reacted cross-linker molecules lower, but the observed behaviour is more complex and points to the importance of the shielding effect. At higher concentrations of the cross-linker there seems to exist a limiting content of doubly reacted units in the copolymer (Fig. 10).

(d) If cyclisation is operative, the relative probability of forming a cycle should depend on the *length and flexibility of the cross-linker bridge*. However, Haward *et al.*[52,54] found an opposite trend in the study of styrene cross-linked with EDMA, tetraoxyethylene dimethacrylate and polyoxyethylene dimethacrylate. Also, the rate of consumption of pendant vinyls

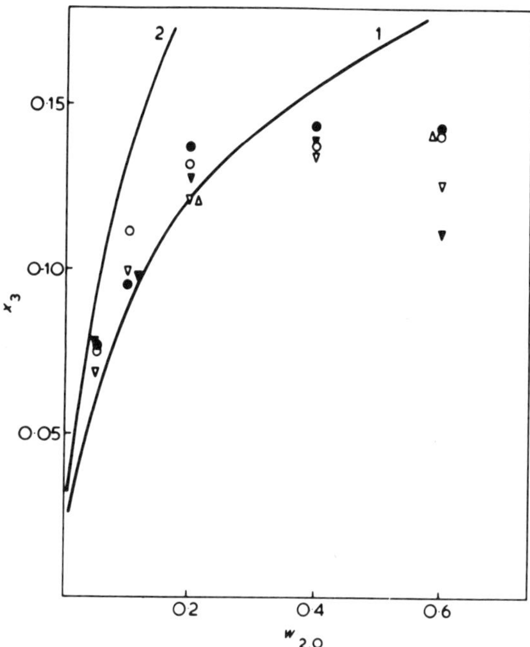

FIG. 10. Dependence of molar fraction of doubly reacted ethylene dimethacrylate (EDMA) units, x_3, in low conversion styrene–EDMA copolymers on the weight fraction of EDMA in feed, $w_{2,0}$.[42] Volume fraction of monomers in feed, v_0: ○ 1·0, ● 0·8, ▽ 0·6, ▼ 0·4, △ 0·2. Curves calculated according to the model of section 2.2.2[42] for $v_0 = 1·0$ (1) and $v_0 = 0·25$ (2).

increased in the same series of cross-linkers; on the contrary, Loshaek and Fox[48] found the fraction of pendant vinyls to increase in the copolymerisation with methyl methacrylate in the series of cross-linkers: EDMA, 2-ethyl-2-butylpropane -1,3-diol and decamethylene-1,10-dioldimethacrylates.

No simple relation could be anticipated between the bridge length and cyclisation probability because, for example, in lactone formation the closure of 8–13-membered rings occurs with a very low probability. The probability of ring closure then increases and the expected decrease obeying approximately the $-3/2$ law occurs at $n > 20$ (e.g. ref. 57). The conformational control of ring formation between terminal groups in linear polymers as well as between side groups on polymer chains has been firmly established (e.g. refs. 30, 58, 59), and it should be valid also for vinyl–divinyl copolymers in the limit of low cross-linker concentrations. At higher concentrations of cross-linker, shielding is progressively operative and the pendant vinyl of a longer and more flexible cross-linker should be less hindered by shielding.

While the independence of, or even a slight decrease in, the fraction $1 - p$ with higher increasing cross-linker concentrations is well established and suggests that some factor (shielding) other than simple ring closure must be involved, the changes at low cross-linker concentrations need further clarification. It is intuitively assumed that the fraction $1 - p$ should increase with increasing x_{DV}^0 at very low x_{DV}^0 values where the chains approach the state of normal coiling and conformational-dependent cyclisation is operative, but as x_{DV}^0 increases further, the shielding effect (to be discussed later in more detail) may progressively become operative and may compensate or overcompensate the tendency of pendant vinyls to participate in cyclisation.

So far, we have considered the composition of the polymer isolated from the polymerisation mixture in the course of polymerisation. Of practical and theoretical interest are results obtained on samples polymerised to the highest possible conversion and sometimes even post-polymerised at a higher temperature to minimise the content of pendant vinyls. In spite of this, pendant vinyls are still present in a considerable quantity, although the conversion by weight reached virtually 100%. In ST–DVB copolymers polymerised with 1% dibenzoyl peroxide at 80°C for 100 h and post-polymerised at 120°C for 50 h, the residual fraction of pendant vinyls determined by i.r. spectrometry changed from 0·2 to 0·3 when the concentration of DVB in the feed increased from 1 to 15 wt %.[60] In ST–EDMA and ST–DVB copolymer beads copolymerised in suspension in the presence of an inert solvent,[61] the fraction p decreased from 0·3 to 0·1 with decreasing cross-linker concentration. From the difference in densities of monomers and the final high conversion copolymer, p was found to vary between 0·5 and 0·2 for copolymers of acrylates and methacrylates with dimethacrylates.[62,63] This result may, however, be somewhat distorted by the fact that the density difference between the monomers and the usually glassy polymers is not equal to the polymerisation contraction measured in solution (melt, rubbery state).

Another possible method of determination of cyclised and singly reacted cross-linker units consists in degradation of soluble or insoluble high conversion cross-linked copolymers by high energy radiation and the study of changes in molecular weight or the amount of sol fraction. Taking a complete inter-molecular reaction of all cross-linker molecules as a reference state, the cross-linking efficiency reduced by both pendant vinyls and cycles could be obtained, amounting to 0·3–0·4.[64,65] However, the calculation depends on a number of approximations that may not be fully justified for this kind of system and irradiation may release further polymerisation of pendant vinyl, if not in the network itself then more easily in the products of scission.

3.2. Efficiency of Cross-linking

In the limit of zero polymerisation conversion, all doubly reacted divinyl units may be regarded as ring forming, because $(1 - p) = (a + s)$ and $a \to 0$ but becomes higher than unity when conversion increases.

The *gel point conversion* is a sensitive measure of the cross-linking efficiency. The cross-linking efficiency α is usually defined by the relation

$$\alpha = a/(a + s) = a/(1 - p) \tag{38}$$

and can be obtained from the ideal (inter-molecular) and experimental cross-linking densities

$$\rho_{id} = \alpha m_\rho/(m_1 + m_2 + m_\rho) \tag{39}$$

$$\rho_{exp} = m_\rho/(m_1 + m_2 + m_\rho) \tag{40}$$

so that

$$\alpha = \rho_{id}/\rho_{exp} \tag{41}$$

Employing eqn. 22 for ρ_{id}, one gets

$$\alpha = [\rho_{exp}(P_w^0 - 1)]^{-1} \tag{42}$$

The value of ρ_{exp} can be obtained experimentally by measuring the content of doubly reacted divinyl units in the copolymer. Equation 42 represents at the same time an implicit relation between the efficiency, α, and critical conversion of vinyls, ξ_c. Assuming the applicability of the ring-free model (section 2.1), eqn. 42 is transformed into

$$\alpha = [\xi_c \Phi_2^2 (P_w^0 - 1)/\varphi_2^0]^{-1} \tag{43}$$

for the copolymerisation of a symmetric independent divinyl monomer, or into

$$\alpha = [\xi_c \varphi_2^0 (P_w^0 - 1)]^{-1} \tag{44}$$

for an ideal copolymerisation.

The part of ρ related to inter-molecular cross-linking (ideal) can also be obtained from the *degree of polymerisation*. The relation to the number-average P_n is straightforward:

$$P_n = P_n^0/(1 - P_n^0 \rho_{id}/2) \qquad (45)$$

since an inter-molecular bond formed reduces the number of molecules by one and $\rho_{id}/2$ is just the number of inter-molecular bonds. By rearrangement, one can linearise eqn. 45 to yield

$$1/P_n^0 - 1/P_n = \rho_{id}/2 \qquad (46)$$

By determining experimentally the fraction of doubly reacted cross-linker molecules, one can obtain α directly.

The determination of α from P_w depends on the assumption of a random distribution of cross-linked units among primary chains. In this case

$$P_w = P_w^0\{1 + P_w^0 \rho_{id}/[1 - (P_w^0 - 1)\rho_{id}]\} \qquad (47)$$

which for $\rho \ll 1$ yields

$$1/P_w^0 - 1/P_w = \rho_{id} \qquad (48)$$

The possible effect of non-random distribution of cross-linked units on P_w and the critical behaviour has been discussed in section 2.1.

From the above equations, it follows that a reliable determination of the degrees of polymerisation of primary chains is important, and, therefore, cross-linker molecules with bridges capable of specific splitting have been used. Wesslau[9] and Mrkvičková and Kratochvíl[53] used bis-4-methacryl-oxybenzylidene ethylenediamine (**I**), which can be split easily by acids to

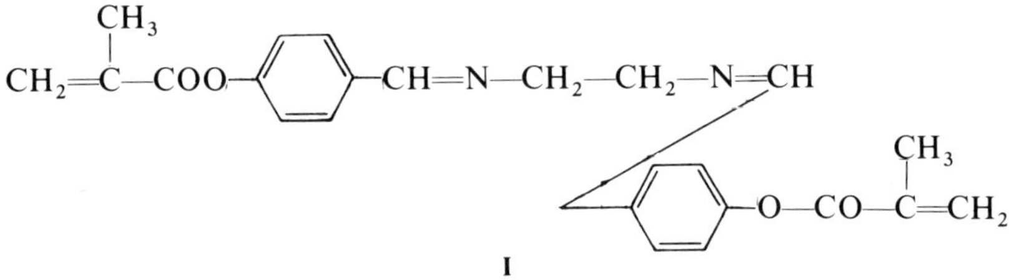

$$CH_2{=}\overset{\displaystyle CH_3}{\underset{\displaystyle |}{C}}{-}COO{-}\langle\bigcirc\rangle{-}CH{=}N{-}CH_2{-}CH_2{-}N{=}CH{-}\langle\bigcirc\rangle{-}O{-}CO{-}\overset{\displaystyle CH_3}{\underset{\displaystyle |}{C}}{=}CH_2$$

I

yield aldehyde groups. These groups were transformed to 2,4-dinitro-phenylhydrazones which show a strong absorption band at 380 nm. Minnema and Staverman[8] used the hydrolysable N,N'-bis-β-methacryl-oxyethyl-p-xylylene diurethane (**II**) and Braun and Brendlein[66,67] copolymerised styrene with 1,4-bis(4-vinylphenyl)-1,2-ethanediol (**III**), in which the α-glycol group can easily be split with lead tetra-acetate. The allyl esters

$$CH_2=\underset{\underset{\displaystyle COO-CH_2-CH_2NHCOOCH_2}{|}}{\overset{\overset{\displaystyle CH_3}{|}}{C}}$$

$$\text{(benzene ring)}-CH_2OCONHCH_2CH_2OCOC(CH_3)=CH_2$$

II

$$CH_2=CH-\text{(benzene ring)}-\underset{\underset{\displaystyle OH}{|}}{CH}-\underset{\underset{\displaystyle OH}{|}}{CH}-\text{(benzene ring)}-CH=CH_2$$

III

can also be hydrolysed, and the average degrees of polymerisation of primary chains can be determined.

The cross-linking efficiency, α, obtained from gel-point conversion measurements by Wesslau[9] varied between 0·4 and 0·1 in the copolymerisation of low concentrations of the degradable cross-linker (0·065–0·75 % based on vinyls) with styrene in the presence of dodecyl mercaptan. The stiff cross-linker cannot cyclopolymerise at all, so that the wastage of 60–90 % cross-links in cycles is due to long-range cyclisation.

Apparently unexpected gel-point conversions were obtained for copolymers of styrene and divinylbenzene[14,51] (Table 2). At a higher concentration of the cross-linker, the critical conversion did not decrease as it should according to eqn. 23, but remained constant or passed through a minimum. For DVB polymerisation, α fell to 0·02–0·03, whereas for 1 % DVB it was 0·2–0·4. Because of the higher reactivity of p-DVB and its higher concentration in the copolymer, α appears to be lower in the copolymerisation of p-DVB than in that of m-DVB. A similar weak dependence of the critical conversion on the cross-linker concentration was reported for styrene–dimethacrylate systems.[54]

The cross-linking efficiency calculated from the critical conversion for poly(N-ethylmethacrylamide)[68] and poly(N-butylmethacrylamide)[69] cross-linked with methylene-bis-acrylamide (MBA) decreased with increasing cross-linker concentration and was lower for the butyl derivative: 0·16, 0·10 and 0·07 (butyl) and 0·36, 0·21 and 0·14 (ethyl) in 75 % solution for 0·26, 0·60 and 1·2 mol % MBA, respectively. The lower efficiency of the butyl polymer can partly be explained by a lower volume concentration of vinyls making cyclisation more probable.

KAREL DUŠEK

TABLE 2
GEL-POINT CONVERSIONS IN THE COPOLYMERISATION OF
STYRENE WITH DIVINYLBENZENE (DVB)

c_{DVB} (wt %)	ξ_{wc} (%)	c_{DVB} (wt %)	ξ_{wc} (%)
p-DVB[14]		*p-DVB*[14]	
0·5	16·1	35	4·5
1	14·1	50	4·5
2·5	7·6	65	5·1
4	7·4	75	5·0
10	5·2	100	4·9
20	4·5		
p-DVB[51]		*m-DVB*[51]	
1	13·2	1	14·1
10	4·4	10	4·9
20	3·5	20	3·3
50	3·6	50	3·6
70	3·6	70	3·4
100	3·6	100	3·3

ξ_{wc} is the critical weight conversion.

Braun and Brendlein[67] determined the number-average molecular weight M_n together with M_n^0 after splitting the cross-linker. The value of α was calculated using eqns. 46 and 41. Figure 11 shows that α is of the order of 10^{-1}, increases with conversion and extrapolates to a value close to zero at zero conversion, as required by the theory. At a given conversion, α depends on dilution; so does gel-point conversion. Thus, for 5 mol % of the cross-linker the conversion by weight of the polymer increased from 0·209 to 0·614, if the polymer concentration decreased from 4 to 1 mol litre^{-1}. This shift of the gel-point conversion was of course partly due to a decrease in P_w^0, but nevertheless α decreased from 0·2 to 0·08. A similar decrease with dilution could also be traced for other concentrations of the cross-linker. No marked dependence of α on the concentration of the cross-linker in the range 1–5 % has been observed.

Mrkvičková and Kratochvíl[53] found, for the copolymerisation of the cross-linker **I** with methyl methacrylate, α to be even lower (0·03–0·3). They calculated the value of ρ_{id} from molecular weights obtained by light scattering using eqn. 48.

The cross-linking efficiency has also been obtained from P_n for diallyl ester polymers. The recalculated data of Matsumoto et al.[45] are given in

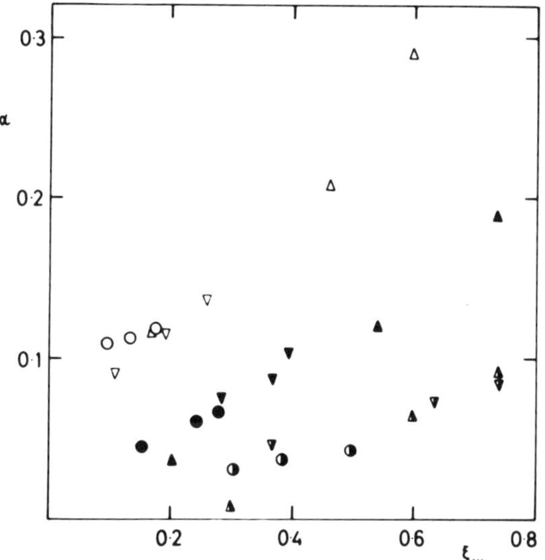

FIG. 11. The cross-linking efficiency in the copolymerisation of 1,4-bis(4-vinyl)ethane-1,2-diol with styrene as a function of conversion (data of ref. 67). Concentration of cross-linker: $\bigcirc$, $\bullet$, $\pmb{\mathbb{O}}$ 5%; $\triangledown$, $\blacktriangledown$, $\pmb{\blacktriangledown}$ 3%; $\triangle$, $\blacktriangle$, $\pmb{\blacktriangle}$ 1%; concentration of monomers: $\bigcirc$, $\triangledown$, $\triangle$ 4 mol litre^{-1}; $\bullet$, $\blacktriangledown$, $\blacktriangle$ 2 mol litre^{-1}; $\pmb{\mathbb{O}}$, $\pmb{\blacktriangledown}$, $\pmb{\blacktriangle}$ 1 mol litre^{-1}.

Table 1. Again, α is rather low, decreases with conversion, and tends to zero when the polymerisation conversion falls to zero.

It can be concluded that the cross-linking efficiency in the pre-gel region is of the order of 10^{-1}–10^{-2}, decreases with decreasing conversion and tends eventually to zero (cf. especially data of refs. 45 and 67). Dilution has a marked effect on α. Therefore an overwhelming majority of doubly reacted divinyl units have been wasted in intra-molecular cross-links.

3.3. Structure and Physical Properties of Pre-gel (Co)polymers

Extensive cyclisation (internal cross-linking) has a marked effect on the shape of polymer molecules and mobility of the segments. It is generally believed that cyclisation greatly reduces the radius of gyration of polymer coils, but it need not be so in all cases. If only small cycles are formed (short-range cyclisation), the chain gets stiffer, resembling partly a ladder chain, and the radius of gyration may increase. However, long-range cyclisation is responsible for a decrease in coil dimensions.

Intrinsic viscosity, $[\eta]$, is a sensitive measure of coil dimensions and reflects well the structural features of polymers obtained by chain (co)polymerisation. A strong reduction of $[\eta]$ with respect to both linear and randomly branched polymers of the same molecular weight is the general observation.

Even for ring-free branched polymers, however, the problem of relating the intrinsic viscosity to their structure is an extremely difficult one and cannot be considered as adequately solved.[70,71] The ratio of intrinsic viscosities of branched (b) and linear (l) polymers of the same molecular weight

$$g' = [\eta]_b/[\eta]_l \tag{49}$$

is smaller than unity and is related to the ratio of unperturbed mean-square radii of gyration

$$g_0 = \langle s_0^2 \rangle_b / \langle s_0^2 \rangle_l \tag{50}$$

as

$$g' = g_0^m \tag{51}$$

where m is a branched structure sensitive parameter which may vary between 1/2 and 3/2 depending on the type of branching.[70] The $\langle s_0^2 \rangle_b$ values are available from experiment or theory, and in this way g' can be used for the determination of the degree of branching, or for a prediction of $[\eta]_b$, if $\langle s_0^2 \rangle_b$ is calculated from the theory (cf. for example refs. 72–74 for random branching).

If branching (cross-linking) is accompanied by cyclisation, which is the usual case in vinyl–divinyl copolymerisation, g' is still smaller than for pure branching. No applicable theoretical model relating g' to the polymer structure exists; nevertheless, vinyl–divinyl copolymers prepared in solution, which must have contained rings, were examined in terms of the ring-free theories (e.g. ref. 75).

Here, we should briefly discuss two kinds of manifestation of branched and cyclised structures: (a) the interrelation between $[\eta]$ and molecular weight and (b) changes in $[\eta]$ occurring during the polymerisation process.

The intrinsic viscosity of polymers with extremely low concentration of cross-linking agent approaches that of the corresponding linear polymer, irrespective of the fraction of cross-links engaged in cycles. If the concentration of the cross-linking agent and dilution increases, the coils become more internally cross-linked and compact and the behaviour of their solutions approaches that of microgel solutions (obtained by the emulsion polymerisation technique) and in the limit that of a suspension of hard spheres, obeying the Einstein law.

A decrease in $[\eta]$ of low conversion ST–DVB copolymers with increasing concentration of DVB was reported by Haward and Simpson[76] as early as 1955.

The results of Zimm *et al.*[77] and Whitney and Burchard[22] are shown in Fig. 12: the intrinsic viscosities are much lower than for the linear polymer and the exponent a in the $[\eta] = KM^a$ dependence is low. However, one must not forget that the structure changes with M, because conversion or dilution change, too. The $[\eta]$ vs. M dependence is very similar to that for microgels obtained by emulsion polymerisation[78] (Fig. 13) and becomes flatter with increasing degree of cross-linking of the microgel particles, approaching[79] the limit of a hard sphere suspension. This similarity suggests that the polymers obtained by cross-linking chain copolymerisation at a higher cross-linker concentration have structures similar to internally cross-linked microgel particles.

The approach of the vinyl–divinyl copolymers to microgel behaviour is very well documented by a steep decrease in $[\eta]_b/[\eta]_l$, extrapolated to zero conversion, with increasing amount of the doubly reacted cross-linker molecules in the copolymer. At 0.7% of doubly reacted divinyl units, $[\eta]$ fell to about 5% of the value found in the absence of the cross-linker.

Also for low conversion soluble EDMA polymers, $[\eta]$ was very low[46] (about 1–5.7% of the value for PMMA) and an analysis showed that $[\eta]$ is

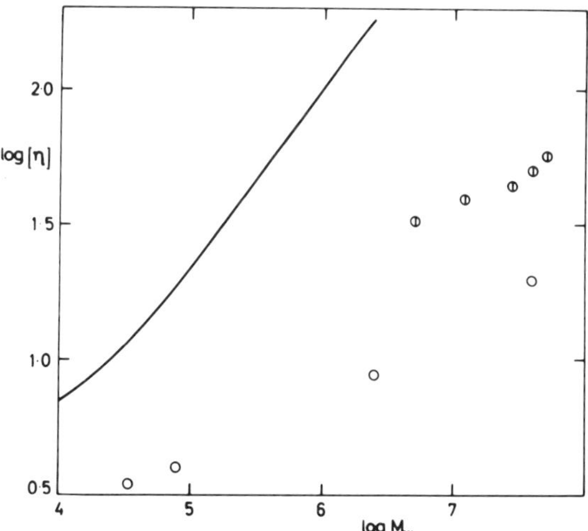

FIG. 12. Dependence of the intrinsic viscosity on molecular weight of copolymers. ○ styrene $+ 10\%$ divinylbenzene, 5–13% solution,[77] $[\eta]$ in toluene; ⊕ pentaerythritol tetramethacrylate–methyl methacrylate,[22] $[\eta]$ in acetone; ——, $[\eta]$ of poly(methyl methacrylate) in acetone.

KAREL DUŠEK

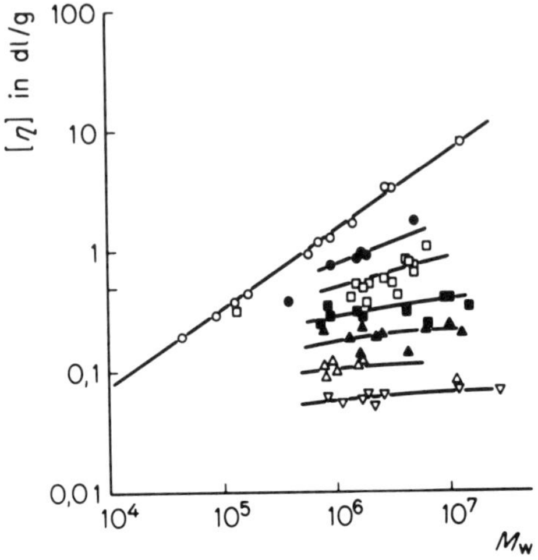

FIG. 13. Intrinsic viscosity–molecular weight relationship for emulsion polymerised microgels from styrene and divinylbenzene (DVB).[78] Solutions in dimethylformamide: ○ linear polystyrene; the other curves from top to the bottom correspond to increasing concentration of DVB.

still much lower than would be expected if only branching were operative. The analysis pointed to a very compact structure of low conversion EDMA polymers.

The dependence of $[\eta]$ on conversion is more complex because an increase in the molecular weight is accompanied by an increase in the degree of branching and cyclisation. For statistically branched ring-free polymers, $[\eta]_\theta$ of theta solutions is related to the number-average radius of gyration and molecular weight:[22]

$$[\eta]_\theta = \Phi(\langle s^2 \rangle^{3/2})_n / M_n \tag{52}$$

where Φ is the Flory's constant. The r.h.s. ratio is much closer to the number-average than to the weight-average and Dobson and Gordon[80] predicted that for random branching $[\eta]$ should increase but remain at the gel-point finite and low (this however refers to $[\eta]$ and not to the bulk viscosity, which diverges at the gel-point!). A similar reasoning should also apply to branching with cyclisation—the difference consists in a higher compactness of the macromolecules and delayed gel-point. Indeed, Whitney and Burchard[22] found for MMA–pentaerythritol tetramethacrylate $[\eta]$ to increase only from 30 to 50 ml g^{-1} if the distance from the

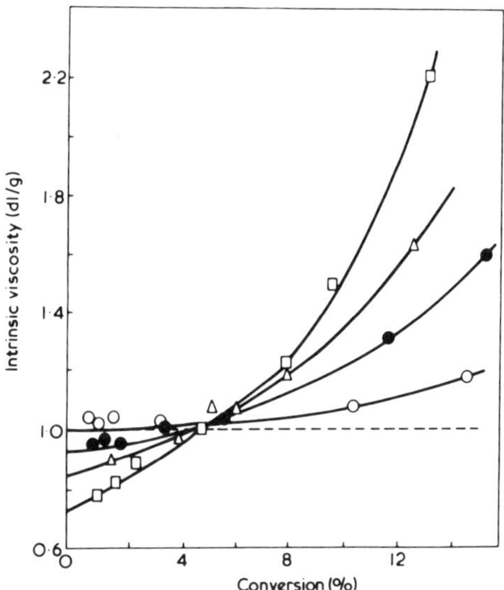

FIG. 14. Dependence of intrinsic viscosity on conversion for styrene–ethylene dimethacrylate (EDMA) copolymers.[52] EDMA in feed (mol %): ○ 1, ● 2, △ 3, □ 4; ---- linear PST.

gel-point, measured by $1 - \xi/\xi_c$, decreased from 5 to 0·25 %; P_w increased by a factor of 30 in the same interval!

Typical changes of $[\eta]$ of ST–EDMA copolymers with increasing conversion were presented by Shah *et al.*:[52] the low conversion value of $[\eta]$ is the smaller and the increase of $[\eta]$ with increasing conversion is the steeper, the higher the cross-linker content (Fig. 14). The zero conversion decrease is due to cyclisation, as has been already discussed (cf. also ref. 55), and the difference in the high conversion parts of curves seems to be primarily associated with a different distance from the gel-point.

However, Ambler and McIntyre[81] found an initial decrease of $[\eta]$ with increasing conversion in the bulk copolymerisation of ST with 0·01–2% DVB and explained this decrease by the effect of ring closure. Thus, both the initial increases or decrease of $[\eta]$ with increasing ξ seem possible, depending on composition of the system and reaction conditions.

It may be concluded that internal cross-linking occurring during cross-linking chain copolymerisation strongly affects the intrinsic viscosity of the polymers formed: $[\eta]$ is much lower than for linear or corresponding branched polymers and its increase with increasing conversion is much slower than would correspond to the increase in molecular weight (there may even occur a temporary decrease of $[\eta]$ with conversion).

The compactness of internally cross-linked molecules should result in a decreased mobility of segments and should be detectable by physical methods. Indeed, the *n.m.r. behaviour* of CCl_4 solutions of low conversion ST–EDMA copolymers gives strong support of the assumption of the compact structure of the copolymers.[82] A fraction of the monomer units of the copolymer does not show directly in high resolution ^{1}H n.m.r. spectra, in contrast to linear ST–MMA or very lightly cross-linked ST–EDMA copolymers as follows from the integrated intensities of n.m.r. bands. The decrease in the intensity was related to the fraction of monomer units sterically hindered (immobilised) in the compact cores of the internally cross-linked molecules. This fraction increases with increasing EDMA content from 0·25 at 12 mol% EDMA in the copolymer to 0·7 at 47 mol% EDMA. It is assumed that the mobile (visible in n.m.r. spectra) part of segments is situated at the periphery of these compact particles. In ^{13}C n.m.r. spectra, however, all monomer units are detected, which means that the segmental motion is sufficiently rapid, but it is incompletely averaged over the space angle. Many swollen cross-linked networks behave in a similar way. Thus, the n.m.r. behaviour points again to the compactness of copolymer molecules, but it makes it possible to distinguish between the hindered core and more mobile periphery of copolymer molecules.

The investigation of *fluorescence depolarisation* by labelled and swollen PST networks[83] in the course of cross-linking was explained by a three-phase structure. The compact particles internally cross-linked are composed of a core covered by a shell (second phase) and interconnected by looser chains (third phase).

3.4. Fraction of Effective Cross-links and Mechanical Behaviour of Networks

Equilibrium stress–strain measurements on rubbery networks, possibly combined with swelling measurements, offer a measure of the fraction of cross-links engaged in elastically active network chains. The connection between the experiment and the structural parameters is possible using a statistical theory of rubber elasticity. Certain experimental conditions and structural requirements which were recently discussed in more detail elsewhere,[84] have to be met for the statistical theory to be applicable. Under these conditions, the experimental equilibrium shear modulus, G_e, is proportional to the volume concentration of elastically active network chains, ν_e, and is composed of covalent (chemical), G_{ec}, and trapped entanglement, G_{ee}, contributions. For the sake of simplicity, we consider here only undiluted and unswollen networks; the modification of the G_{ec} term to include these effects is possible (cf. refs. 84, 85).

$$G_e = G_{ec} + G_{ee} = AR T v_{ec} + eR T T_{eg} \tag{53}$$

where R is the gas constant and T is temperature, A is the front factor in the statistical theory of rubber elasticity, e is a constant and T_{eg} is the so-called trapping factor approaching unity for a perfect network. The value of the chemical contribution to v_e, v_{ec}, as well as of T_{eg} is available, e.g. from the theory of branching processes (the so-called chain-end correction is automatically included, cf. for example ref. 23).

The determination of the fraction of effective cross-links, ε, is based on the following reasoning: sufficiently far beyond the gel-point (for example, if the cross-linking density exceeds the value at the gel-point by a factor of 10) and for a high degree of polymerisation of primary chains, the sol fraction becomes negligible, all cross-links except those in elastically inactive cycles become elastically active, and all potential entanglements have become trapped ($T_{eg} \approx 1$), i.e. G_{ee} is constant. The increase in v_{ec} is then equal to the increase in the concentration of double bonds of the cross-linker in feed, v, after the concentrations corresponding to divinyl units with pendant vinyls and engaged in elastically inactive cycles have been subtracted. Therefore, the incorporation of an additional cross-linker molecule (of two cross-linked units) in the network will contribute less than two effectively cross-linked units. In this region, the increase in v_e, Δv_e, is related to ε:

$$\varepsilon = \Delta v_e / \Delta v = (\Delta G_e / \Delta v)/AR T \tag{54}$$

The situation is visualised in Fig. 15, where ε is the slope of the linear part of the curve. The determination is based on the assumption of the negligible chain-end effect, constancy of entanglement contribution and of the cyclisation and shielding factors. The linear dependence of G_e vs. v may be taken as an indication that these requirements have been fulfilled. In a broader range of v, the G_e vs. v dependence curves are usually curved upwards[60] and the departures from Gaussian elasticity behaviour may be the main reason.

According to eqn. 54, determination of ε from G_e requires a knowledge of the value of the front factor A. The figures given below are all related to the front factor $A = 1$ which is valid for a phantom network with fully suppressed fluctuation of cross-links.

Wesslau[86] examined in this way copolymers of MMA with various

                    KAREL DUŠEK

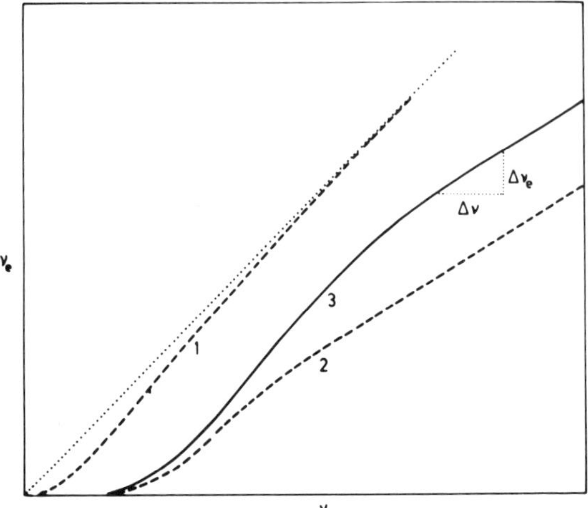

FIG. 15. Schematic representation of the dependence of the concentration of elastically active network chains, v_e, on the concentration of cross-linked units, v, and of the cyclisation and trapped entanglement effects. (1) Ring-free chemical contribution to v_e, (2) taking into account formation of elastically inactive cycles and shielding, (3) contribution 2 + trapped entanglement contribution.

dimethacrylates; ε increased in the series methacrylic anhydride (0·37), butane-1,4-diol dimethacrylate (0·60) hydroquinone dimethacrylate (0·78) and 4,4′-dihydroxydiphenyl dimethacrylate (0·89). The decrease in ε can be well correlated with increasing tendency to cyclisation. It is very plausible that the decrease in ε is mainly caused by cyclisation, because at the level of cross-linking used ($v \approx 5 \times 10^{-4}\,\text{mol ml}^{-1}$) the fraction of singly reacted cross-linker molecules is expected to be low. That cyclisation has the main effect on ε is corroborated by the decrease in ε with dilution, e.g. in 2-hydroxyethylmethacrylate copolymers with EDMA[87] $\varepsilon = 0·7$ for bulk polymerisation and 0·5 for 50 % diluent. The value of ε depends not only on the structure of the cross-linker, but also on the flexibility of the main chain and volume concentration of pendant vinyls affecting the level of the inter-molecular cross-linking. The decrease in ε in the series 0·70, 0·54, 0·36, 0·40 for cross-linked polymers of methyl, ethyl, butyl and octyl acrylates, respectively, has been explained[88] just by the decreasing volume concentration of vinyls. In poly(octyl acrylates), partial ordering of alkyl side chains was detected, which may have been the reason for the somewhat higher value of ε. A lower volume concentration of vinyls may also be the reason for a lower value of $\varepsilon = 0·43$ for N-butylmethacrylamide networks[69] cross-linked with methylene bisacrylamide, as compared with $\varepsilon = 0·83$ for

N-ethylmethacrylamide networks.[68] The higher ε observed for polymethacrylate than polyacrylate networks[89] may be only apparent and caused by the tendency to self-cross-linking of polyacrylates due to chain transfer, released by abstraction of hydrogen on the alpha carbon atom.

In general, ε varies between 0·4 and 0·9 and seems to be determined primarily by the fraction of elastically inactive cycles. These figures have been obtained, however, under the assumption that the front factor is $A = 1$. If $A = 1/2$, valid for phantom networks with tetrafunctional junctions, ε would exceed unity, which does not seem physically realistic.

There exists a relation between the density, conversion of pendant vinyls and *modulus of rigidity* of highly cross-linked and *glassy polymer networks*. Application of high pressure favours further polymerisation of pendant vinyls; as a result, the density is higher after release of the pressure, and so is the modulus of rigidity.[90,91]

A view has been put forward that the modulus can be related to the spatial density of covalent bonds so that a polymer with high modulus could be prepared if it were possible to build up an extensive covalent bond structure. By increasing the polymerisation pressure to 2·5 GPa, it was possible to increase the density of poly(p-divinylbenzene)[89] from 1·05 to 1·18 g ml^{-1} and decrease the fraction of pendant double bonds to 0·35. At the same time, the modulus of rigidity increased by a factor of two, to 4 GPa. Still higher density of bonds can be obtained by pressure polymerisation of 1,3,5,-trivinylbenzene.[91] At lower temperatures, there was little difference between the densities of poly(p-divinylbenzene) and poly(trivinylbenzene), but at 250 °C trivinyl benzene gave a higher density product. An increase in pressure caused again a decrease in the concentration of pendant vinyls. The highest moduli of the products were higher by a factor of 4 compared with the modulus of polystyrene. It is interesting to note that none of the materials showed evidence of a glass transition below 230 °C and that the specific heats decreased with increasing polymer density and moved in towards the level associated with a diamond structure.

4. FORMATION AND STRUCTURE OF HETEROGENEOUS (MACROPOROUS) NETWORKS

4.1. Main Features of Macroporous Network Formation and Structure

Cross-linking chain (co)polymerisation is a suitable method of preparation of macroporous networks used as materials for sorption or chromatographic separation, as catalysts or matrices for ion exchange resins. The features of this process summarised in section 5 and refs. 92–95 play certainly an important role.

If polyvinyl compounds are polymerised or copolymerised with monovinyl monomers in the presence of an inert (non-polymerisable) diluent, heterogeneous networks may be formed, which may contain after removal of the solvent microvoids (macropores). The term macroporosity is usually connected with dry porosity characterised by a lower density of the network due to voids than that of the matrix polymer.* However, there exist situations when, for example, the network shows up distinct regions separated by well developed boundaries in the swollen state, but in the dry state the density is equal to the density of the matrix polymer. At low concentrations of monomers, a coherent network is no longer formed and a solution of a microgel is the final product, resembling microgels prepared by emulsion copolymerisation of vinyl and multivinyl monomers. Such systems have been extensively studied by Funke (refs. 98–100 and papers quoted therein) and others.[78,79,101,102]

The literature sources pertaining to macroporous networks, their modification and application, are very numerous, but here we will limit ourselves with a short overview of the preparation–structure relationships as far as they are connected with the mechanism of cross-linking (co)polymerisation. Attention has been paid especially to cross-linked polystyrenes (styrene–divinylbenzene copolymers, ST–DVB); the main features of preparation, structure and properties of these networks based on results obtained till 1966 were summarised in a review.[103] The interest in this system still predominates although other macroporous systems are also important (cf., for example hydrophilic polymers[104–107]). Newer results on ST–DVB networks were presented in refs. 95–98, 105, 108–127.

The main factors governing the formation of heterogeneous (macroporous) networks can be summarised as follows:

(1) Heterogeneous or macroporous structure is obtained only above a certain threshold concentration of the diluent and cross-linking agent (cf. Fig. 16 for ST–DVB systems).
(2) At a given concentration of the cross-linker, the threshold con-

---

* Nomenclature in the field of polymer sorbents and their classification have been a subject of discussion.[95–97] In contrast to this definition of macroporosity by existence of voids of any size, which lower the density of the sample, some authors reserve the term macroporosity for voids larger than 25 nm and range smaller voids with micropores. However, microporosity is sometimes identified with the so-called gel porosity, which exists in the swollen state, and is used for characterisation of the permeability of the gel.

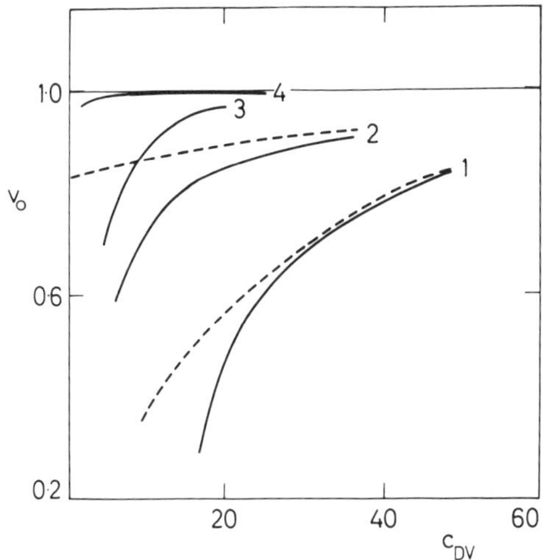

FIG. 16. Threshold volume fraction of the diluent, $1 - v_0$, and concentration of divinylbenzene, c_{DV} (wt. %), in the mixture with styrene necessary for the formation of heterogeneous copolymers.[105] ---- Onset of turbidity in non-extracted samples; ——— onset of permanent porosity in extracted dry samples.

centration of the diluent depends on its thermodynamic interaction with the network polymer and other components of the system (the monomers, etc.) (Fig. 16). Heterogeneous and macroporous structures are obtained not only with non-solvating diluents (like aliphatic hydrocarbons or alcohols for the ST–DVB systems), but also with good solvents (e.g. toluene). At a given concentration of the diluent, the threshold concentration of the cross-linker decreases with deteriorating thermodynamic quality of the diluent. (3) In the homologous series of diluents, the threshold concentration of the diluent strongly depends on the size of the diluent molecule (its molar volume). Higher members of the series are more effective (aliphatic hydrocarbons,[112,115,116,119,125] esters[127] or alcohols[115]). Polystyrenes having chains composed of the same units as network chains are effective in producing macroporous structures but only beyond a certain limit of molecular weight. Polymers with dissimilar monomer units, which are incompatible with the matrix polymer, are the most powerful diluents (e.g. polydimethylsiloxanes for ST–DVB networks[105,125]).
(4) If the concentration of diluent and cross-linker is below the threshold value, apparently homogeneous (transparent) networks are formed after polymerisation, extraction and drying. Their density is equal to the density

of the matrix polymer. The inhomogeneity (cross-linking density fluctuations) characteristic for cross-linking chain copolymerisation may be enhanced by dilution, however.

(5) The increase in porosity (pore volume) after passing the threshold concentration of the cross-linker and diluent is initially rapid. With increasing concentration of the cross-linker, the dry porosity, p_d, soon reaches a constant value; with increasing concentration of the diluent it continues to increase or reaches a constant value.

(6) The specific surface area of macropores measured in the dry state after extraction by gas adsorption or mercury porosimetry depends to a great extent on the nature and amount of diluent: polymers produce the largest, and well solvating diluents the smallest, pores. For ST–DVB copolymers, typical values of the specific surface area are $5\,m^2\,g^{-1}$ for polystyrenes, $200\,m^2\,g^{-1}$ for C_7–C_8 hydrocarbons and $500\,m^2\,g^{-1}$ for the good solvent toluene.[129] Aliphatic alcohols usually produce larger pores than aliphatic hydrocarbons, presumably because of a difference between the interfacial energies of the diluent and network polymer. The specific surface area usually passes through a maximum with increasing concentration of the diluent.[108,112] At high dilutions especially with non-solvating diluents, the pore structure is coarse and the network loses its coherence.

Points (1)–(6) give only a crude restatement of regularities characteristic of the formation of heterogeneous cross-linked structures; specifications of particular systems as well as of combination of diluents are described in the pertinent literature referred to above.

4.2. Formation Mechanism

It could be assumed that the formation of heterogeneous and macroporous networks structures is exclusively determined by features of the chain cross-linking (co)polymerisation and proceeds as follows: the microgel-like particles (nuclei) originally formed are tied up together by chains containing much fewer cross-linker molecules. These less cross-linked domains are much more swollen with the diluent–monomer matrix and eventually by the diluent. Upon drying, the structure cannot fully collapse and the formed macropores correspond to interparticle spaces. The change in the thermodynamic quality of the diluent is reflected in a different degree of association of the nuclei.[128]

However, there are strong indications that the inhomogeneous character of chain cross-linking polymerisation is not a sufficient condition for the formation of two-phase structures, where the second phase is either the diluent after polymerisation or voids after drying, and that a phase separation mechanism is operative. The importance of phase separation seems now to have been accepted.[95,103,106,114,125] According to this approach, the system separates into two phases as soon as its limit of thermodynamic stability is passed; phase separation can take place on a macroscale (macrosyneresis, deswelling) or on a microscale (microsyneresis), which is typical for the formation of disperse two-phase structures including macroporous networks.[105]

The following arguments can serve to support the importance of a phase separation mechanism: the sudden appearance of heterogeneity or macroporosity, when the threshold concentrations of the cross-linker and diluent have been passed; the usually much larger dimensions of the porous structure than those of primary nuclei; a great influence of the size and nature of diluent molecules; and the fact that disperse porous structures can also be obtained by other cross-linking mechanisms like cross-linking of primary chains in solution (e.g. refs. 130, 131).

A thermodynamic theory was developed with the assumption of phase equilibrium on a microscale in every step of the polymerisation reaction, based on Flory's statistical thermodynamic theory of swelling and taking into account changes in the network structure in the course of the reaction.[103,105,125,126,130] Phase separation sets in when the Gibbs energy of the homogeneous mixture becomes higher than the sum of Gibbs energies of the two phases which grow from the initial composition. The approach towards the thermodynamic stability limit is stimulated not only by worsening of the thermodynamic quality of the monomer–diluent mixture (mixing term), but also by an increase in the elastic term, which is proportional to the cross-linking density and molar volume of the diluent. In order to get a permanently heterogeneous system, phase separation must occur in the course of the cross-linking process so that the disperse structure is fixed by formation of new cross-links.

It is the feature of cross-linking copolymerisation that a network is formed at relatively low conversions and is therefore highly swollen with the diluent–monomer mixture, so that phase separation commences not far beyond the gel-point.

The network phase is usually, but not always, smaller in volume than the liquid phase. The change in the volume of the separated network (gel) phase is controlled by several factors—by thermodynamic interactions determined by the monomer–diluent ratio, increase in the cross-linking density and transfer of monomer from the liquid phase into the gel phase in the course of polymerisation. Such paradoxes as the higher volume of pores than is the volume of the diluent or the lower volume degree of swelling than is the value obtained for the corresponding undiluted system of the same composition,[103,105] can be explained on the basis of this approach.[102,125] Experiments show that the greater the pore size, the sooner phase separation starts with respect to the gel-point. If the separation occurs before the gel-point, the structure is either very coarse and incoherent (concentrated polymer phase is the minority phase) or it contains relatively large droplets (diluent phase is the minority phase, e.g. the case for polydimethylsiloxane diluents) due to coalescence before the structure is fixed by the network of cross-links.

The agreement between the experiments and predictions of the thermodynamic theory is relatively good,[105,125] but the theory should be applied with care because it disregards the inhomogeneous structure of the system before phase separation. However, an extension of the phase separation theory to inhomogeneous systems would be very difficult.

On the other hand, phase separation occurring when the network density is higher (which is the case for cross-linking of solutions of linear polymers) results in a fine pore structure. So, polystyrene cross-linked to a high degree in a 9% solution in a relatively good solvent yields porous networks with a very high specific surface area amounting to $1000\,\mathrm{m}^2\,\mathrm{g}^{-1}$ and small pores 5–7 nm in size (seen by small angle X-ray scattering[131,132]). Due to the small pore size, the networks are transparent and due to a high cross-linking density of the matrix they swell almost to the same degree in all solvents irrespective of their thermodynamic interaction with polystyrene.

The additional question should be discussed, whether phase separation starts in the metastable region close to the equilibrium limit (binodal)[133] (i.e. in the vicinity of composition corresponding to coexisting phases, by nucleation and growth) or if the changes occurring (increase in cross-linking density, change in the composition, etc.) are so rapid that phase separation occurs, when the limit of absolutely unstable region (described as the spinodal) has been reached, by a spinodal decomposition mechanism.[134–137] In this limit, the system is unstable to small fluctuations which grow in amplitude and form originally a uniformly disperse structure. This mechanism occurs in metal alloys and glasses and sometimes in concentrated solutions of some polymers.[137] A large undercooling (e.g. ~50 K) is typical for this mechanism. However, this mechanism is valid only for a short time as the system spontaneously tends to reach the equilibrium between separated phases which is (at least in solutions) accompanied by structural reorganisation (coarsening due to coalescence). It has not been investigated to what extent this mechanism could apply to phase separation during cross-linking chain polymerisation. It does not seem to be operative in mobile systems separated from solution or weak gels i.e. in the vicinity of the gel-point. At least, the undercooling found in temperature induced phase separation in gels[138] is by one or two orders of magnitude smaller than that observed for the spinodal decomposition mechanism. However, such a mechanism is more probable for phase separation in rigid structures like those described in refs. 131 and 132.

The generally inhomogeneous course of the chain cross-linking (co)polymerisation is reflected in the complex structure of the separated network phase. For instance, if precipitants are used as diluents, the dimensions of the resulting structural elements are of the order of 10^2 nm. A closer electron microscopic examination reveals, however, a finer structure of the polymer phase responsible for the high specific surface and presumably composed of primary nuclei or their aggregates.[111]

4.3. Effect of Aftertreatment on Structure

The dry porosity does not necessarily correspond to the situation after network formation. Several copolymers that are turbid and exhibit a two-phase structure after polymerisation become transparent, one-phase and non-porous after removal of the solvent by drying. On the other hand, one-phase transparent swollen polymer gels become turbid and exhibit a two-phase behaviour after the (good) swelling agent is replaced by a poorer solvent. This procedure is well known in the preparation of porous membranes (cf. for example ref. 139). Both observations mean that the porous structure can be modified, created or destroyed by changes in physical conditions.

The disappearance of porosity after drying and its reappearance after reswelling[97,108,109,112,119,124,128,140] is due to capillary contraction or collapse of pores. It has been found that during solvation (hydration in the case of ion exchange resins), the cryptoheterogeneous polymer with collapsed pores absorbs the solvent first and only after it becomes sufficiently flexible do the pores start opening.[108] The most susceptible to collapse and reopening are fine pores and this is the reason that the size of the pores (specific surface area) decreases upon drying. The main reason for the pore collapse can be seen in a change in the interfacial energy and cohesion of pore walls that have come close together. Since pore collapse and reopening is accompanied by deformation of the matrix, the process is easier, the lower the cross-linking density of the matrix. So collapse has not been observed upon drying in highly cross-linked systems where the fine pore structure remains intact.

However, the collapse and reopening phenomena depend on the way the diluent has been removed upon drying. The collapse is typical for evaporation of the diluent or solvent well above the glass transition temperature of the system.

The large difference between the 'porosity' in the dry and swollen samples can be documented by small angle X-ray scattering[123] measurements on a ST–DVB copolymer prepared in the presence of octane: the dry porosity is close to zero while the volume fraction of the diluent phase in the copolymer swollen in toluene is 0·48.

If, however, the solvent has been replaced by a poorer one, the pores do not collapse and the porosity remains at least partially preserved. By a series of solvent exchanges, one can pass from good solvents (e.g. toluene in the case of ST–DVB copolymers) to completely non-solvating liquids (like water).[124]

The changes induced by a solvent exchange in porous materials are intimately connected with the formation of a pore structure in networks that are originally apparently homogeneous and swollen. The solvent exchange produces a heterogeneity which is often accompanied by appearance of turbidity and is analogous to the situation when the solvent power of the swelling agent in a swollen gel is lowered, e.g. by a decrease in temperature.[105,138] Slow relaxation of the network as a whole is the reason why the excess solvent is not expelled out of the sample to form a new bulk phase (deswelling) and why a fine dispersion is formed instead. A phase equilibrium mechanism seems to be operative,[105] but spinodal decomposition has also been suggested.[141] It is important that this type of phase separation is stronger and easier, the weaker the network, in contrast to the formation of chemically fixed heterogeneity occurring during cross-linking copolymerisation, which is stronger the higher the degree of cross-linking. Such a physically induced transient heterogeneous structure can be converted into a 'permanent' porous structure upon removal of the solvent below the glass transition temperature of the matrix and fixed by additional cross-linking (e.g. by radiation) if necessary.

4.4. Characterisation of Network Structure of the Matrix Polymer

If the expansion of the macroporous structure on swelling occurs isotropically, the degree of equilibrium swelling can be used for determination of the concentration of elastically active network chains in the matrix polymer. However, the front factor A in the elasticity theory (cf.

eqn. 53) is a function of the network history, i.e. it depends on relative volumes of the network phase at which the cross-links have been formed. Because this volume changes in the course of copolymerisation, the resulting networks can be regarded as composite,[130] sometimes even expanded with A higher than the value for an undiluted network.[126] Without information about the history of porous structure formation, no assumption can be made about the value of the front factor A. The determination of A is possible from swelling data in different solvents provided the polymer–solvent interaction parameter for the respective pairs is known.[123] A quantitative interpretation of the equilibrium swelling is impossible, if the condition of isotropy of swelling of any part of the structure is violated.

The determination of the network parameters from the bulk modulus of the porous systems in a rubbery state (either at sufficiently high temperature or swollen) should be considered with great care, because the simple geometry of the bulk strain (e.g. extension or compression) is complicated by bending and buckling of pore walls.[124]

5.   CONCLUSIONS

5.1. Structure of (Co)polymers

The experimental findings regarding the structure and behaviour of the (co)polymers will be summarised and compared with the predictions of the ring-free models.

The function $1 - p$ of *doubly reacted cross-linker units* in the copolymer at low conversions is unexpectedly high. For pure bis-unsaturated monomers, it amounts to 60 % for diallyl phthalate, 50 % for p-DVB and 30 % for EDMA. There is only a weak increase of $1 - p$ with increasing conversion and dilution. A similar behaviour was observed for vinyl–divinyl copolymers, but the conversion and dilution dependences were stronger at low concentrations of the cross-linker. At higher concentrations of the divinyl monomer, however, the divinyl molecules additionally introduced into the copolymer by increasing their fraction in the feed are built into the structure only as units with pendant vinyls. When the polymerisation reaction has reached completion, a considerable fraction of units with pendant vinyl remains in the copolymer even at a polymerisation temperature well above the glass transition temperature. The ring-free models predict a gradual decrease in p from unity at zero conversion to $p = 0$. Therefore, two main factors must be operative: the first raises, and the second reduces, the apparent reactivity of pendant vinyls. It can be seen that both factors are interdependent: the initial activation of the pendant vinyls brings about their blockage when a sufficient number of units have doubly reacted. These factors have been identified as cyclisation (internal or intra-molecular cross-linking) and shielding of the pendant vinyls in the internally cross-linked structure. There exist additional arguments for both factors; viz., decrease in p with increasing cyclisation probability or increase in the apparent reactivity of pendant vinyls with increasing length and flexibility of the cross-linker bridge (reduction in shielding). This view is corroborated by other observations.

The *efficiency of cross-linking*, α, expressed as the fraction of doubly reacted bis-unsaturated units, obtained at low conversions from molecular weight or gel-point conversion measurements, amounts to 10^{-1} at low cross-linker concentrations and falls to 10^{-2} at higher concentrations. The fraction α falls to zero at zero conversion, and decreases with increasing dilution and increasing cross-linker concentration. That only a small fraction of bis-unsaturated units is engaged in inter-molecular cross-links is a proof of the importance of the cyclisation reaction.

The *intrinsic viscosity* of pre-gel (co)polymers is in general low and departure from the behaviour of linear polymers increases with increasing cross-linker content, conversion and dilution. The increase in $[\eta]$ with conversion is relatively weak, because an increase in the molecular weight is compensated for by an increase in compactness of the molecules. The intrinsic viscosities are considerably lower than values that would correspond to randomly branched acyclic molecules and, at a higher cross-linker content, approach the behaviour of microgels obtained by cross-linking polymerisation in emulsion. Thus, the viscosity behaviour is an indication of a compact structure of the copolymer molecules brought about by internal cross-linking. Further support for the existence of compact structures in copolymers has been obtained from the analysis of *^1H n.m.r. spectra* of pre-gel copolymers. A part of the compact particles (presumably their core) is so hindered that the respective protons do not show in the high resolution spectra. The fraction of the units in the strongly cross-linked core increases with increasing concentration of the cross-linker. The other part shown in the spectra is presumably situated at the particle periphery.

In this connection, the term 'shielding' used for the first time by Minnema and Staverman[8] deserves a comment. By shielding we understand a decrease in the ability of a pendant vinyl, buried inside the internally cross-linked structure, to react, should it be due to its own low mobility (diffusion control) or imposed steric hindrance, or to the inability of a macroradical to propagate inside such a structure.

5.2. Formation of Copolymers and Networks

The important role of cyclisation and its double effect on the apparent reactivity of pendant vinyls, structure of the copolymers and gelation was recognised a long time ago and restated recently more comprehensively.[22,23,47,92] In a recent paper, however, it was argued that cyclisation does not play any major role and that it is solely a decreased reactivity of the pendant vinyl which is responsible for the deviations.[12]

We will consider here two extreme cases of chain cross-linking copolymerisation:

(a)   very low concentration of the cross-linker, and
(b)   high concentration of the cross-linker.

In case (a), a fraction of cross-linker units reacts inter-molecularly and a fraction intra-molecularly, because the growing macroradical has in every propagation step a finite chance to attack the double bond pendant on its own or on an attached chain; however, the ring closure probability decreases with the separation between the radical and pendant double bond. Because there are very few divinyl units in the chain, there are very few rings in chains, all pendant vinyls are almost equally accessible and the polymer coils approach in the limit the coils of linear chains. Models of copolymerisation kinetics based on the mass action law are applicable which work with the average concentration of pendant vinyls in the system, as far as inter-molecular cross-linking is concerned (cf. section 2).

In case (b), i.e. at a high concentration of the divinyl monomer, network formation proceeds approximately as shown in Fig. 17: primarily, compact internally cross-linked molecules resembling microgel particles are formed which are linked together with participation of their peripheral vinyls via a growing macroradical in the 'solution' phase. Dilution promotes intra-particle contacts: the growing macroradical has a better chance to coil back and attack another peripheral vinyl on the same particle. This is why the gel-points are shifted on dilution to higher conversion. Cyclisation measured by the concentration of intra-molecular cross-links and contributed mainly by wasted internal cross-links in the core increases only little, because dilution has a minor effect on the core structure.

The heterogeneous course of the chain cross-linking polymerisation and existence of regions where propagation cannot occur (forbidden pockets) which appear to be identical with internally cross-linked cores, has been

                           KAREL DUŠEK

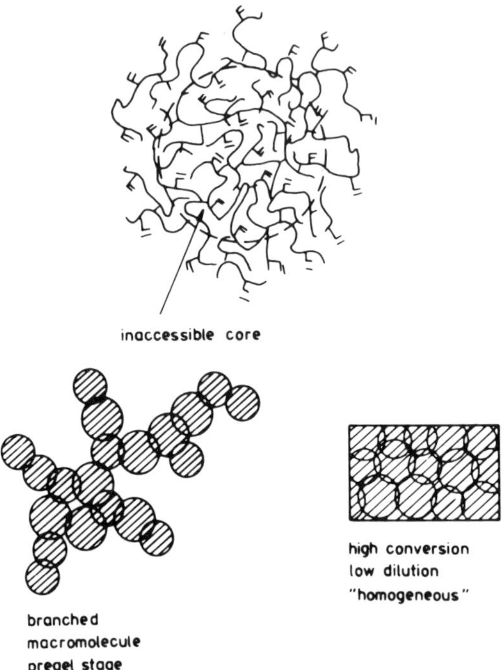

FIG. 17.   Mechanism of network formation in chain cross-linking (co)polymerisation at higher concentrations of the cross-linking agent (taken from ref. 92).

established by Notley.[93] Also, Roshchupkin *et al.* obtained evidence of the microgel character of polymers formed from diacrylates.[94]

For this reason, in the cross-linker-rich region the copolymerisation kinetic models based on the average concentration of pendant vinyls are not applicable as far as additions of the growing radical are considered. Also, the application of cyclisation probability based on the conformational statistics of single connecting paths (e.g. spanning-tree approximation) is incorrect with regard to the dominating correlations due to cycles.

Any statistical treatment should take into account the fact that the microgel-like particles can be linked together inter-molecularly only through the vinyls in their surface layers. Assuming an approximately constant thickness of the surface layer, the fraction of bonds engaged in inter-molecular bonds will be lower, the larger the molecule. In the polydisperse system, this means a decrease in the apparent reactivity of pendant vinyls in larger molecules in relation to smaller ones, a factor which should further delay gelation on the conversion scale.

5.3. Structure of Networks

The high conversion networks obtained using a low concentration of the cross-linker resemble the usual networks with a few dangling ends and pendant double bonds and some cross-links wasted in elastically inactive cycles. The equilibrium stress–strain behaviour of such swollen networks usually corresponds to that predicted for Gaussian networks. The fraction of effective cross-links in most cases varies between 0·5 and 0·8; the ineffective part is lost mainly in cycles.

The situation is different at medium and high cross-linker concentrations. The heterogeneous structure of microgel-like particles suspended in monomer solutions must persist beyond the gel-point. Paradoxically, the high conversion polymers do not show any distinct heterogeneity. However, the inter-particle space is gradually filled in with the newly formed polymer from the monomer with participation of peripheral vinyls of the already existing particles, and the density fluctuations become more or less smoothed out (Fig. 17).

If a diluent is present during polymerisation, a permanently heterogeneous network may be formed, possibly containing pores after drying. Not only poor solvents causing precipitation of the corresponding linear polymer are effective diluents, but also good solvents and polymers composed of segments identical with segments of the network chains (e.g. linear polystyrene for styrene–divinylbenzene copolymers). Porosity appears only above a certain threshold concentration of the diluent and cross-linker. This threshold concentration is lower, the lower the solvent power of the diluent and the larger its molecules.

The tendency towards the inhomogeneous course of cross-linking chain copolymerisation is of importance for pore formation, but phase separation occurring in the course of copolymerisation seems to be a necessary condition. The porous structure, characterised by the total pore volume, specific surface area and pore size distribution, also depends on aftertreatment. Pores may disappear upon drying by a capillary contraction mechanism and new pores may be generated as a result of solvent exchange by a microsyneresis mechanism, if a good solvent in the swollen gel has been replaced by a poor solvent.

NOTE

Since the time of submission of this chapter a number of papers on cross-linking chain copolymerisation and copolymers have been published. No substantial breakthrough in the theory, structural characterisation or application has occurred, but the author feels that some amending facts and interpretations may make the picture more complete.

The dependence of the apparent reactivity of the pendant double bond on the length of the bridge and on the overall degree of cross-linking was demonstrated by O'Donnell and O'Sullivan[142] and by Krakovyak *et al.*[143] Polymerisation of a long bis-unsaturated monomer 4,6,9,12,14-penta-oxa-1,16-heptadecadiene-5,13-dione (structure **IV**) used for manufacturing

$$CH_2{=}CH{-}CH_2{-}O{-}CO{-}O{-}CH_2{-}CH_2{-}O{-}CH_2$$
$$CH_2{-}O{-}CO{-}O{-}CH_2{-}CH{=}CH_2$$

(**IV**)

optical lenses, yielded high conversion polymers with only about 5% unreacted double bonds.[142] Similarly, practically the same reactivity of double bonds in a divinyl monomer and of the resulting pendant vinyls was observed in copolymerisation with only 0·02–0·2% divinyl monomer, where no immobilisation or shielding due to cyclisation is expected because of an extremely low concentration of the cross-linker.[143] Further experimental support for the importance of cyclisation in vinyl–divinyl copolymerisation was obtained by Galina and Rupicz,[144] the cross-linking efficiency obtained from the gel-point conversions at different dilutions of styrene and ethylene dimethacrylate monomers was of the order of 10^{-1}. The authors also offer a modification of the theory of branching processes to include mild cyclisation into the cross-linking copolymerisation treatment. The application of the fluorescence polarisation method to changes in the structure in the course of the copolymerisation of styrene with divinylbenzene[145] revealed structural inhomogeneities, which did not change to any great extent when the gel-point was passed. With increasing cross-linker concentration, the segmental density increased and pendant double bonds became less accessible. The older and newer arguments were recently summarised and analysed and the decisive role of cyclisation was pointed out.[146] On the contrary, Hild and Rempp[147] offered a reasoning based mainly on the data of ref. 12, suggesting that the special features of cross-linking chain copolymerisation were due to a decreased reactivity of the pendant double bonds which was, however, not caused by cyclisation.

Of interest to experimentalists may be the progress made in methods suitable for analysis of products, i.e. reassessment of i.r. spectroscopy,[148] application of Raman spectrometry,[142] fluorescence depolarisation for characterising segmental density and structural inhomogeneity[145] and luminescence methods for analysis of the composition of the copolymer containing a labelled cross-linker.[143]

Studies of formation, structure and properties of heterogeneous (macroporous) copolymers continue to be very numerous and are stimulated by the variety of applications of these copolymers. The isoporous polystyrene matrices, obtained by cross-linking polystyrene chains in solution[131,132] were found to be efficient sorbents for organic substances from aqueous solutions.[149] The pore structure was found to be dependent on the nature of the solvent used: with decreasing solvent power, the total porosity increased and larger pores were formed.[150] Also, new data on the changes of the pore structure generated by solvent exchange in the matrix have been presented.[151]

Swatling et al.[152] investigated in detail the morphology of membranes obtained by copolymerisation of styrene and divinylbenzene in the presence of various diluents including paraffin wax. The polymerisation took place between glass plates. With the exception of the wax, the authors always observed formation of a non-porous skin which is either smooth or crenulated and which affects the transport properties of the membrane. The polymerisation contraction and filling of the resulting gap at the mould surface with a monomer–diluent mixture containing less cross-linker (due to its faster consumption in copolymerisation) may be the reason. The concentration of DVB in the skin layer may fall below the critical value so that the porous structure can no longer be formed. Interfacial phenomena may be another reason.

Solution copolymerisation of mono- and divinyl monomers at higher concentrations of the cross-linker and/or dilution produces insoluble particulate material of microgel type and has been called *particulate precipitation polymerisation*.[153] Addition of a diluent of a lower solvating power for the copolymer appeared to optimise the reaction medium for the formation of particles of a desired size. Various reactive groups can be incorporated into the matrix by the choice of the monovinyl comonomer. The application of these powdery products as material for chromatographic separation, polymeric reagents, supports for the Merrifield type synthesis of polypeptides, segments, or foaming agents for plastics has been suggested.[152–158]

REFERENCES

1. Dušek, K., Gordon, M. and Ross-Murphy, S. B., *Macromolecules*, **11** (1978), p. 236.
2. Flory, P. J., *Principles of Polymer Chemistry*, Ithaca, New York, Cornell Univ. Press, 1953.

3. STOCKMAYER, W. H., *J. Chem. Phys.*, **11** (1943), p. 45.
4. GORDON, M. and SCANTLEBURY, G. R., *Trans. Faraday Soc.*, **60** (1964), p. 604.
5. BUTLER, G. B., *J. Polym. Sci.*, *Polymer Symp.*, **64** (1978), p. 71.
6. ALFREY, T., JR, BOHRER, J. J. and MARK, H., *Copolymerization*, New York, Interscience, 1952.
7. GIBBS, W. E., *J. Polym. Sci.*, *Part A*, **2** (1964), p. 4809.
8. MINNEMA, L. and STAVERMAN, A. J., *J. Polym. Sci.*, **29** (1958), p. 281.
9. WESSLAU, H., *Angew. Makromol. Chem.*, **1** (1967), p. 56.
10. DUŠEK, K., *Collect. Czech. Chem. Commun.*, **34** (1969), p. 1891.
11. BRAUN, D. and BRENDLEIN, W., *Makromol. Chem.*, **167** (1973), p. 203.
12. OKASHA, R., HILD, G. and REMPP, P., *Eur. Polym. J.*, **15** (1979), p. 975.
13. VALVASSORI, A. and SARTORI, G., *Adv. Polym. Sci.*, **5** (1967), p. 28.
14. STOREY, B. T., *J. Polym. Sci.*, *Part A*, **3** (1965), p. 265.
15. MALINSKÝ, J., KLABAN, J. and DUŠEK, K., *Collect. Czech. Chem. Commun.*, **34** (1969), p. 711.
16. WILEY, R. H., *Pure Appl. Chem.*, **43** (1975), p. 57.
17. SCHWACHULA, G., *J. Polym. Sci.*, *Polym. Symp.*, **53** (1974), p. 107.
18. POPOV, G. and SCHWACHULA, G., *Plast. Kautsch.*, **24** (1979), p. 374.
19. TIDWELL, P. W. and MORTIMER, G. A., *J. Macromol. Sci.*, *Revs. Macromol. Chem.*, **C4** (1970), p. 281.
20. DUŠEK, K., *Makromol. Chem.*, Suppl. 2 (1979), p. 35.
21. KUCHANOV, S. I. and PISMEN, L. M., *Vysokomol. Soedin.*, **A13** (1971), p. 2035.
22. WHITNEY, R. S. and BURCHARD, W., *Makromol. Chem.*, **181** (1980), p. 869.
23. STOCKMAYER, W. H., *J. Chem. Phys.*, **12** (1944), p. 125.
24. LOWRY, G. G., *Markov Chains and Monte-Carlo Simulations in Polymer Science*, M. Dekker, New York, 1970.
25. HOLT, T. and SIMPSON, W., *Proc. Roy. Soc. (London)*, **A238** (1956), p. 154.
26. ASO, C., *J. Polym. Sci.*, **39** (1959), p. 475.
27. ROOVERS, J. and SMETS, G., *Makromol. Chem.*, **60** (1963), p. 89.
28. MATSUMOTO, A., ASO, T., TANAKA, S. and OIWA, M., *J. Polym. Sci.*, *Polym. Chem.*, **11** (1973), p. 2357.
29. JACOBSON, H. and STOCKMAYER, W. H., *J. Chem. Phys.*, **18** (1950), p. 160.
30. SEMLYEN, T., *Adv. Polym. Sci.*, **21** (1976), p. 41.
31. KIRKPATRIC, S., *Revs. Mod. Phys.*, **45** (1973), p. 574.
32. STAUFFER, D., *J. Chem. Soc.*, *Faraday Trans. 2*, **72** (1976), p. 1354.
33. STAUFFER, D., Lecture at the 7th Discussion Conference *Polymer Networks*, Carlsbad, Czechoslovakia, 1980, to be published in *Pure Appl. Chem.*
34. GORDON, M. and SCANTLEBURY, G. R., *Proc. Roy. Soc. (London)*, **A292** (1966), p. 1380.
35. GORDON, M. and SCANTLEBURY, G. R., *J. Chem. Soc. B* (1967), p. 1.
36. GORDON, M. and SCANTLEBURY, G. R., *J. Polym. Sci.*, *Part C*, **16** (1968), p. 3933.
37. GORDON, M., TEMPLE, B., PARKER, T. G. and LOVE, A. J., *J. Prakt. Chem.*, **313** (1971), p. 411.
38. STEPTO, R. F. T., Chapter 3.
39. HAWARD, R. N., *J. Polym. Sci.*, **14** (1954), p. 535.
40. DUŠEK, K. and ILAVSKÝ, M., *J. Polym. Sci.*, *Polym. Symp.*, **53** (1975), p. 57.
41. DUŠEK, K. and ILAVSKÝ, M., *J. Polym. Sci.*, *Polym. Symp.*, **53** (1975), p. 75.

42. DUŠEK, K. and SPĚVÁČEK, J., *Polymer*, **21** (1980), p. 750.
43. SIMPSON, W., HOLT, T. and ZETIE, R. J., *J. Polym. Sci.*, **10** (1953), p. 489.
44. ITO, K., MURASE, Y. and YAMASHITA, Y., *J. Polym. Sci., Polym. Chem.*, **13** (1975), p. 87.
45. MATSUMOTO, A., YOKOYAMA, S., KHONO, T. and OIWA, M., *J. Polym. Sci., Polym. Phys.*, **15** (1977), p. 127.
46. GALINA, H., DUŠEK, K., BOHDANECKÝ, M., ŠTOKR, J. and TUZAR, Z., *Eur. Polym. J.*, **16** (1980), p. 1043.
47. KAST, H. and FUNKE, W., *Makromol. Chem.*, **180** (1979), p. 1335.
48. LOSHAEK, S. and FOX, T. G., *J. Amer. Chem. Soc.*, **75** (1953), p. 3544.
49. COSTA, L., CHIANTORE, O. and GUAITA, M., *Polymer*, **19** (1978), p. 198.
50. TSUKINO, M. and KUNITAKE, T., *Macromolecules*, **12** (1979), p. 387.
51. MALINSKÝ, J., KLABAN, J. and DUŠEK, K., *J. Macromol. Sci., Chem.*, **A5** (1971), p. 1071.
52. SHAH, A., HOLDAWAY, I., PARSONS, I. W. and HAWARD, R. N., *Polymer*, **19** (1978), p. 1067.
53. MRKVIČKOVÁ, L. and KRATOCHVÍL, P., *J. Polym. Sci., Polym. Phys.*, in press.
54. SHAH, A., PARSON, I. N. and HAWARD, R. N., *Polymer*, **21** (1980), p. 825.
55. SOPER, B., HAWARD, R. N. and WHITE, E. F. T., *J. Polym. Sci., Part A-1*, **10** (1972), p. 2542.
56. ESCHWEY, H. and BURCHARD, W., *J. Polym. Sci., Polym. Symp.*, **53** (1974), p. 1.
57. SISIDO, M., *Macromolecules*, **4** (1971), p. 6.
58. IMANISHI, Y., *J. Polym. Sci., Macromol. Revs.*, **41** (1979), p. 1.
59. PLATÉ, N. A. and NOAH, O. V., *Adv. Polym. Sci.*, **31** (1979), p. 133.
60. DUŠEK, K., *Collect. Czech. Chem. Commun.*, **27** (1962), p. 2841.
61. ŠTOKR, J., SCHNEIDER, B., FRYDRYCHOVÁ, A. and ČOUPEK, J., *J. Appl. Polym. Sci.*, **23** (1979), p. 3553.
62. KATZ, D. and TOBOLSKY, A. D., *J. Polym. Sci.*, **A2** (1964), p. 1593.
63. TOBOLSKY, A. V., KATZ, D., TAKAHASHI, M. and SCHAFFHAUSER, R., *J. Polym. Sci.*, **A2** (1964), p. 2749.
64. SHULTZ, A. R., *J. Amer. Chem. Soc.*, **80** (1958), p. 1854.
65. HWA, J. C. H., *J. Polym. Sci.*, **58** (1963), p. 715.
66. BRAUN, D. and BRENDLEIN, W., *Makromol. Chem.*, **146** (1971), p. 117.
67. BRAUN, D. and BRENDLEIN, W., *Makromol. Chem.*, **167** (1973), p. 217.
68. ULBRICH, K., ILAVSKÝ, M., DUŠEK, K. and KOPEČEK, J., *Eur. Polym. J.*, **13** (1977), p. 579.
69. ULBRICH, K., DUŠEK, K., ILAVSKÝ, M. and KOPEČEK, J., *Eur. Polym. J.*, **14** (1978), p. 45.
70. SMALL, P. A., *Adv. Polym. Sci.*, **18** (1975), p. 1.
71. BOHDANECKÝ, M. and KOVÁŘ, J., *Viscosity of Polymer Solutions*, Amsterdam, Elsevier, 1981.
72. ZIMM, B. H. and STOCKMAYER, W. H., *J. Chem. Phys.*, **17** (1949), p. 1301.
73. GORDON, M. and MALCOLM, G. N., *Proc. Roy. Soc. (London)*, **A295** (1966), p. 29.
74. KAJIWARA, K. and GORDON, M., *J. Chem. Phys.*, **59** (1973), p. 3626.
75. KAMADA, K. and SATO, H., *Polym. J.*, **2** (1971), p. 489.
76. HAWARD, R. N. and SIMPSON, W., *J. Polym. Sci.*, **18** (1955), p. 440.

77. ZIMM, B. H., PRICE, F. P. and BIANCHI, J. P., *J. Phys. Chem.*, **62** (1958), p. 979.
78. HOFFMAN, M., *Makromol. Chem.*, **175** (1974), p. 613.
79. SHASHOUA, V. E. and BEAMAN, R. G., *J. Polym. Sci.*, **33** (1958), p. 101.
80. DOBSON, G. R. and GORDON, M., *J. Chem. Phys.*, **41** (1964), p. 2389.
81. AMBLER, M. R. and MCINTYRE, D., *J. Appl. Polym. Sci.*, **21** (1977), pp. 237, 2269, 3237.
82. SPĚVÁČEK, J. and DUŠEK, K., *J. Polym. Sci., Polym. Phys.* (1980).
83. FUHRMANN, J. and LEICHT, R., *Colloid Polym. Sci.*, **258** (1980), p. 631.
84. DUŠEK, K. and ILAVSKÝ, M., *Polym. Eng. Sci.*, **19** (1979), p. 246.
85. DUŠEK, K. and PRINS, W., *Adv. Polym. Sci.*, **6** (1969), p. 1.
86. WESSLAU, H., *Makromol. Chem.*, **93** (1966), p. 55.
87. HASA, J., ILAVSKÝ, M. and JANÁČEK, J., *Collect. Czech. Chem. Commun.*, **32** (1967), p. 4156.
88. ILAVSKÝ, M., HASA, J. and DUŠEK, K., *J. Polym. Sci., Polym. Symp.*, **53** (1975), p. 239.
89. JANÁČEK, J. and FERRY, J. D., *J. Polym. Sci., Part A-2*, **10** (1972), p. 345.
90. RACKLEY, F. A., TURNER, H. S., WALL, W. F. and HAWARD, R. N., *J. Polym. Sci., Polym. Phys.*, **12** (1974), p. 1355.
91. PRICE, L., HAWARD, R. N. and PARSONS, I. W., *Polymer*, **20** (1979), p. 162.
92. DUŠEK, K., GALINA, H. and MIKEŠ, J., *Polym. Bull.*, **3** (1980), p. 19.
93. NOTLEY, N. T., *J. Chem. Phys.*, **66** (1966), p. 1577.
94. ROSHCHUPKIN, V. P., OZERKOVSKII, V. B., KALMYKOV, YU. B. and KOROLEV, G. V., *Vysokomol. Soedin*, **A19** (1977), p. 699.
95. TAGER, A. A. and TSILIPOTKINA, M. V., *Usp. Khim.*, **47** (1978), p. 152.
96. MILLAR, J. R., *J. Polym. Sci., Polym. Symp.*, **68** (1980), p. 167.
97. SEDEREL, W. L. and DE JONG, G. J., *J. Appl. Polym. Sci.*, **17** (1973), p. 2835.
98. FUNKE, W., BEER, W. and SEITZ, U., *Prog. Colloid Polym. Sci.*, **57** (1975), p. 48.
99. OBRECHT, W., SEITZ, U. and FUNKE, W., *Makromol. Chem.*, **177** (1976), p. 2235.
100. HILLER, J.-CH. and FUNKE, W., *Angew. Makromol. Chem.*, **76/77** (1979), p. 161.
101. PRICE, C., FORGET, J.-L. and BOOTH, C., *Polymer*, **18** (1977), p. 526.
102. KÄMMERER, H.-F. and MÜCK, K. F., *Macromol. Chem.*, **136** (1970), p. 161.
103. SEIDL, J., MALINSKÝ, J., DUŠEK, K. and HEITZ, W., *Adv. Polym. Sci.*, **5** (1967), p. 113.
104. WICHTERLE, O., 'Hydrogels' in *Encyclopedia of Polymer Science*, Volume 15, New York, Interscience, 1971.
105. DUŠEK, K., in *Polymer Networks. Structure and Mechanical Properties*, Chompff, A. J. and Newman, S. (Eds.), New York, Plenum Press, 1971.
106. ČOUPEK, J., KŘIVÁKOVÁ, M. and POKORNÝ, S., *J. Polym. Sci., Part C*, **42** (1973), p. 185.
107. DAWKINS, J. V. and GABBOTT, N. P., *Polymer*, **22** (1981), p. 291.
108. KUN, K. A. and KUNIN, R., *J. Polym. Sci., Part A-1*, **6** (1968), p. 2689.
109. HILGEN, H., DE JONG, G. J. and SEDEREL, W. L., *J. Appl. Polym. Sci.*, **19** (1975), p. 2647.
110. JACOBELLI, H., BARTHOLIN, M. and GUYOT, A., *Angew. Makromol. Chem.*, **80** (1979), p. 31.

111. BARTHOLIN, M. and SPITZ, R., *6th Disc. Conf. Polymer Networks, Carlsbad, Czechoslovakia* (1981), Preprint No. 62.

112. BRUTSKUS, T. K., SALDADZE, K. M., UVAROVA, E. A. and LYUSTGARTEN, E. I., *Kolloid. Zh.*, **34** (1972), p. 509.

113. BRUTSKUS, T. K., SALDADZE, K. M., UVAROVA, E. A. and LYUSTGARTEN, E. I., *Kolloid. Zh.*, **34** (1972), p. 672.

114. BRUTSKUS, T. K., SALDADZE, K. M., FEDTSOVA, M. A., UVAROVA, E. A., LYUSTGARTEN, E. I., ITKINA, M. I. and SLADKAYA, L. D., *Vysokomol. Soedin.*, **17** (1975), p. 1247.

115. TAGER, A. A., TSILIPOTKINA, M. V., MAKOVSKAYA, E. B., LYUSTGARTEN, E. I., PASHKOV, A. B. and LAGUNOVA, M. A., *Vysokomol. Soedin.*, **A13** (1971), p. 2370.

116. TAGER, A. A., TSILIPOTKINA, M. V., MAKOVSKAYA, E. B., PASHKOV, A. B., LYUSTGARTEN, E. I. and PECHENKINA, M. A., *Vysokomol. Soedin.*, **A10** (1968), p. 1065.

117. MIKES, J., *Symp. Kunstharz-Ionenaustauscher, Plenar-Diskussions-Vorträge*, Berlin, Akademie-Verlag, 1970, p. 53.

118. HÄUPKE, K., *Symp. Kunstharz-Ionenaustauscher, Plenar-Diskussions-Vorträge*, Berlin, Akademie-Verlag, 1970, p. 158.

119. HÄUPKE, K. and PIENTKA, V., *J. Chromatogr.*, **102** (1974), p. 117.

120. HÄUPKE, K. and PIENTKA, V., *Proc. 3rd Symp. Ion Exchange*, Balatonfüred, 1974, p. 99.

121. WOLF, F. and SCHAAF, R., *Plaste Kautsch.*, **17** (1970), p. 323.

122. GALINA, H. and KOLARZ, B. N., *Polym. Bull.*, **2** (1980), p. 235.

123. BALDRIAN, J., KOLARZ, B. N. and GALINA, H., *Coll. Czech. Chem. Commun.* (1981), in press.

124. WIECZOREK, P. P., ILAVSKÝ, M., KOLARZ, B. N. and DUŠEK, K., *J. Appl. Polym. Sci.*, in press.

125. DUŠEK, K., SEIDL, J. and MALINSKÝ, J., *Coll. Czech. Chem. Commun.*, **32** (1967), p. 2767.

126. BERANOVÁ, H. and DUŠEK, K., *Coll. Czech. Chem. Commun.*, **34** (1969), p. 2932.

127. VAŠÍČEK, Z., *Proc. 3rd Symp. Ion-Exchange*, Balatonfüred, 1974, p. 393.

128. MILLAR, J. R., SMITH, D. G., MARR, W. E. and KRESSMAN, T. R. E., *J. Chem. Soc., London*, **218** (1963), p. 2779.

129. SEIDL, J., MALINSKÝ, J. and KREJCAR, E., *Chem. Prumysl.*, **15** (1965), p. 414.

130. DUŠEK, K., *J. Polym. Sci., Part C*, **16** (1967), p. 1289.

131. DAVANKOV, V. A., ROGOZHIN, S. V., TSYURUPA, M. P. and PANKRATOV, E. A., *Zh. Fiz. Khim.*, **48** (1974), p. 2964.

132. DAVANKOV, V. A. and TSYURUPA, M. P., *Angew. Makromol. Chem.*, **91** (1980), p. 127.

133. KONINGSVELD, R., *Adv. Colloid Interface Sci.*, **2** (1968), p. 151.

134. CAHN, J. W. and HILLIARD, J. E., *J. Chem. Phys.*, **29** (1958), p. 258.

135. ZARZYCKI, J., *Disc. Faraday Soc.*, **50** (1970), p. 122.

136. VAN AARTSEN, J. J., *Eur. Polym. J.*, **6** (1970), p. 919.

137. SMOLDERS, C. A., VAN AARTSEN, J. J. and STEENBERGEN, A., *Kolloid Z.-Z. Polym.*, **243** (1971), p. 14.

138. DUŠEK, K. and SEDLÁČEK, B., *Coll. Czech. Chem. Commun.*, **34** (1969), p. 136.

139. KESTING, R. E. and MANNEFEE, *Kolloid Z.-Z. Polym.*, **230** (1968), p. 341.
140. OSTRIKOV, M. S., DUKHINA, T. P., VLODAVETS, I. N. and SINITSYNA, G. M., *Kolloid Zh.*, **26** (1964), p. 600.
141. COHEN, C., TANNY, G. B. and PRAGER, S., *J. Polym. Sci., Polym. Phys.*, **17** (1979), p. 477.
142. O'DONNELL, J. H. and O'SULLIVAN, P. W., *Polym. Bull.*, **5** (1981), p. 103.
143. KRAKOVYAK, M., ANANIEVA, T., ANUFRIEVA, E., GOTLIB, Y., NEKRASOVA, T. and SKOROKHODOV, S., *Makromol. Chem.*, **182** (1981), p. 1009.
144. GALINA, H. and RUPICZ, K., *Polym. Bull.*, **3** (1980), p. 473.
145. LEICHT, R. and FUHRMANN, J., *Polym. Bull.*, **4** (1981), p. 141.
146. GALINA, H. and GORDON, M., *Proc. 12th Europhysics Conf. Molecular Mobility in Polymer Systems*, Leipzig, 1981, p. 50.
147. HILD, G. and REMPP, P., *Pure Appl. Chem.*, **53** (1981), p. 1341.
148. BARTHOLIN, M., BOISSIER, G. and DUBOIS, J., *Makromol. Chem.*, **182** (1981), p. 2075.
149. DAVANKOV, V. A., VOLYNSKAYA, A. B. and TSYURUPA, M. P., *Vysokomol. Soedin., Pt. B*, **22** (1980), p. 746.
150. TSYURUPA, M. P., PANKRATOV, E. A. and DAVANKOV, V. A., *Vysokomol. Soedin. Pt. B*, **22** (1980), p. 755.
151. KOLARZ, B. N., WIECZOREK, P. P. and WOJACZYNSKA, M., *Angew. Makromol. Chem.*, **96** (1981), p. 193.
152. SWATLING, D. K., MANSON, J. A., THOMAS, D. A. and SPERLING, L. H., *J. Appl. Polym. Sci.*, **26** (1981), p. 591.
153. BAMFORD, C. H., LEDWITH, A. and SEN GUPTA, P. K., *J. Appl. Polym. Sci.*, **25** (1980), p. 2559.
154. HINTZSCHE, W., HÄUPKE, K., POPOV, G., SCHWEIGER, M. and SCHWACHULA, G., *Plaste Kautsch.*, **28** (1981), p. 248.
155. MIZUTANI, Y. and MATSUOKA, S., *J. Appl. Polym. Sci.*, **26** (1980), p. 2113.
156. MIZUTANI, Y., KUSUKOTO, K. and KAGIYAMA, Y., *J. Appl. Polym. Sci.*, **26** (1981), p. 271.
157. MIZUTANI, Y., MATSUOKA, S. and KUSUMOTO, K., *J. Appl. Polym. Sci.*, **17** (1973), p. 2925.
158. MIZUTANI, Y., MATSUOKA, S., IWASAKI, M. and MIYAZAKI, Y., *J. Appl. Polym. Sci.*, **17** (1973), p. 3651.

Chapter 5

NETWORK FORMATION IN VULCANISATION OF ELASTOMERS

M. MORTON

*Institute of Polymer Science, University of Akron,
Ohio, USA*

SUMMARY

The practical utilisation of elastomers depends strongly on the vulcanisation process, which ties the long-chain molecules into permanently elastic networks by a cross-linking reaction. Although this process originated with Charles Goodyear's discovery of the reaction between sulphur and natural rubber, it is now applied to a variety of elastomers, using various cross-linking agents. Among the polymers included in this discussion are: natural rubber, styrene–butadiene rubber, polybutadiene, nitrile rubber, butyl, ethylene–propylene rubber (EPDM), polychloroprene (neoprene), silicone rubber, polyethylene, chlorbutyl, chlorosulphonated polyethylene, urethane rubbers and fluoropolymers.

As might be expected, the various cross-linking agents can lead to different network structures, and it has been the goal of polymer science to relate these structures to their mechanical and chemical behaviour.

1. INTRODUCTION

Although rubber was first discovered in the New World at the time of Columbus's voyages, it was not really possible to utilise it in any practical way until after 1839, when Charles Goodyear in America and Thomas Hancock in England independently and almost simultaneously discovered the process of sulphur 'vulcanisation'. As the name implies ('Vulcan', god of fire), this was a treatment of rubber at elevated temperatures in the presence of sulphur, with rather remarkable effects on the properties, i.e. the conversion of a somewhat sticky, soft, viscoelastic material into a tough, highly elastic one. Although the nature of the chemical changes caused by this process, i.e. the cross-linking of linear long-chain molecules into a network, was not understood until almost a century later, this process permitted the rapid development of rubber technology for the production of a vast array of products, from rubber bands to rubber tyres.

One of the remarkable facts in modern rubber technology is that, although there are now available a number of alternative practical methods for the formation of rubber networks, sulphur vulcanisation still represents the principal one used today, for both natural and synthetic rubbers, 140 years after its discovery! There have, of course, been vast improvements in carrying out this chemical reaction so that today's process is a far cry from the original primitive methods. But sulphur is still the principal cross-linking agent because no other process can match *all* of the advantageous properties of this material, i.e. low cost, low toxicity, ease of process control and optimum vulcanisate properties.

It is, of course, true that sulphur vulcanisation can only be used for those elastomers which contain some chemical unsaturation, but they still represent the majority of these materials. The others require different chemical agents to accomplish the cross-linking reaction, depending on the chemical constitution of the polymer chains. The most 'universal' of these cross-linking agents are probably the organic peroxides, which can, as a rule, be used for both saturated and unsaturated polymers. This universality of the peroxides is due to the fact that they can be readily decomposed to yield free radicals, which can attack the usually abundant carbon–hydrogen bonds, abstracting hydrogen atoms to leave carbon radicals which readily couple to form cross-links between chains.

Although these organic peroxides can be used to cross-link almost all types of elastomers, there are other preferred methods, especially in the case of polymers containing specific reactive atoms or functional groups in the chain. Table 1 lists the more common methods of cross-linking polymers, principally elastomers, together with the current opinion about the nature of the cross-links obtained. These chemical reactions are discussed in greater detail in subsequent sections.

It may be appropriate at this point to emphasise that the main scientific interest in the chemistry of the cross-linking of macromolecules has been to establish the structures of the resulting networks, since the nature of these

TABLE 1

CROSS-LINKING AGENTS FOR ELASTOMERS

Cross-linking agent	Elastomer	Type of cross-link
Sulphur (accelerated)	Unsaturated polymers:	
	natural rubber	Mono- to polysulphides
	SBR	Mono- to polysulphides
	polybutadiene	Mono- to polysulphides (?)
	polyisoprene	Mono- to polysulphides (?)
	nitrile rubber	Mono- to polysulphides (?)
	butyl	Disulphides
	EPDM	Disulphides (?)
Peroxides	Unsaturated polymers (except butyl)	Carbon–carbon (C—C)
	Saturated polymers: silicones polyethylene urethanes	Carbon–carbon (C—C)
Metal oxides	Halogenated polymers: polychloroprene chlorbutyl chlorosulphonated polyethylene	Metal–carbon|or ether
	Carboxylated polymers	Metal salt
Di- or poly-amines	Halogenated polymers (as above)	Carbon–nitrogen[a]
	Fluoropolymers	Carbon–carbon[a] and carbon–nitrogen
Phenolic resins Quinones Maleimides	Unsaturated polymers	Carbon–carbon (C—C)

[a] Can also be aromatised.

structures can have a profound effect on both their mechanical and chemical behaviour. Unfortunately, this has proved to be a most difficult task because of the following factors:

(a)   The low concentration of cross-links (1–2%) involved in rubber networks. In many cases, much higher concentrations of cross-linking agents have been used with model compounds in order to elucidate the reactions, but such studies do not necessarily permit a valid extrapolation to the low concentrations used in forming the actual network.

(b) The difficulties involved in analysis of a 'non-molecular' species such as a network, i.e. insolubility in any solvent. To 'dissolve' such a network must involve chain scission and degradation which of course can obscure the results.

(c) The complexity of some of the chemical reactions. This is particularly true in the case of sulphur vulcanisation.

2. SULPHUR VULCANISATION

2.1. Natural Rubber

Ever since the development of vulcanisation accelerators early in this century, it was realised that the type of vulcanisation process used could markedly influence the physical and chemical properties of the resulting product. This was particularly true of the aging behaviour. Hence there was intense interest in elucidating the detailed structure of vulcanised rubber in order to correlate it with the observed properties. The early naïve ideas about cross-linking occurring through simple addition of sulphur across double bonds were soon modified by the realisation that many other reactions could and did occur, e.g. substitution, polysulphide formation, cyclisation, double-bond isomerisation, etc. It is not surprising, therefore, that, despite the pioneering and elegant studies by such workers as Armstrong, Little and Doak,[1] Selker and Kemp,[2] Farmer and co-workers,[3,4] Shelton and McDonel,[5] and many others, no definitive picture of sulphur vulcanisates of natural rubber emerged until the more recent work by Bateman and his group of co-workers.[6-10]

It might be instructive, before presenting the current ideas developed by Bateman and colleagues, to describe briefly the methods they used to achieve their results. These were based on a two-pronged approach, as follows:

(a) chemical identification of the products of the reaction between sulphur and model olefins (and diolefins);[6-9]

(b) determination of the stoichiometry of the reaction of sulphur with natural rubber (in presence or absence of accelerators, etc.) in relation to the number of cross-links formed, and chemical identification of types of cross-links (i.e. monosulphide, disulphide, polysulphide, etc.[6,8,10]).

The model olefin mainly used under (a) above was 2-methylbutene-2, corresponding to the monomer unit in natural rubber. However, the use of the diolefin 2,6-dimethyloctadiene-2,5, representing a *double* unit in natural rubber, was even more instructive, since it demonstrated the formation of cyclic sulphides (5-membered and 6-membered) by *intra-molecular* cyclisation reactions. In contrast, *inter-molecular* formation of sulphides would, of course, represent cross-linking. This shows the importance of using the right model compounds, since the single unit butene would not be so capable of showing the occurrence of cyclisation.

The methods used for (b) above consisted of the following:

(i)   Analysis for amount of combined sulphur in the natural rubber vulcanisate.
(ii) Determination of cross-link density (number of cross-links) from elasticity measurements.
(iii) Determination of proportion of polysulphides (and some of the disulphides) by extraction of the vulcanisate with triphenyl-phosphine.
(iv) Cleavage of sulphur–sulphur bonds by treatment with di-*n*-butyl phosphite, lithium aluminium hydride, or thiol–amine mixtures.[11]

The 'chemical probes' listed under (iii) and (iv) above are of special interest. Thus the triphenylphosphine has the special property of 'shortening' a polysulphide chain by successively removing sulphur atoms as follows:

$$(C_6H_5)_3P + RS_xR \longrightarrow (C_6H_5)_3PS + RS_{x-1}R \longrightarrow \cdots \qquad (1)$$

This 'stripping' of sulphur atoms continues, in the presence of sufficient phosphine, until the sulphide link is reduced to a *disulphide*, when neither of the R groups is alkenyl, or to a *monosulphide* when at least one of the R groups is alkenyl. Hence the amount of sulphur removed from the vulcanisate is a measure of the polysulphidic content of the cross-links.

Similarly the di-*n*-butyl phosphite can be used to degrade the vulcanisate by cleaving the sulphur–sulphur cross-links. The residual network can then be considered as held together by monosulphidic cross-links, or carbon–carbon bonds. All of these reactions are, of course, performed on the swollen vulcanisate in an inert solvent.

A comment is also in order about step (ii) above, which involves a determination of the cross-link density from elasticity measurements. The basic rubber elasticity equation is given by:

$$f/A_0 = \rho g R T M_c^{-1}(\alpha - \alpha^{-2})(1 - 2M_c M_n^{-1}) \qquad (2)$$

TABLE 2[6]

VULCANISATION OF NATURAL RUBBER WITH 10 phr[a] SULPHUR AT 140°C

Vulcanisation time (h)	Moles cross-links per g rubber ($\times 10^5$)	Combined sulphur (%)	S atoms consumed per cross-link formed	S removed by $(C_6H_5)_3P$ (%)	No. of S atoms in each cross-link	S in cyclic sulphides (%)
2	1·0	1·68	53	20	12–13	76–77
4	2·0	3·46	53	16	10–11	79–81
7	4·2	5·93	47	13·5	7–8	83–85
24	7·1	8·89	43	2·5	2–3	93–95

[a] phr = parts by weight per hundred of rubber.

TABLE 3[6]

ACCELERATED[a] SULPHUR VULCANISATION OF NATURAL RUBBER AT 100°C

Vulcanisation time (h)	Moles cross-links per g rubber ($\times 10^5$)	Combined sulphur (%)	S atoms consumed per cross-link formed	S removed by $(C_6H_5)_3P$ (%)	No. of S atoms in each cross-link
3	1·07	0·28	8·2	61	3·2
24	9·4	0·83	2·8	32	1·9
144	16·2	0·99	1·9	18	1·6

[a] See Table 4 for vulcanisation recipe.

where $f =$ force of extension

$\quad A_0 =$ cross-sectional area of *unextended* sample

$\quad \rho =$ density of rubber

$\quad \alpha =$ extension ratio (i.e. stretched length/unstretched length)

$\quad M_c =$ molecular weight between cross-links, i.e. number-average molecular weight of the network chains

$\quad M_n =$ number-average molecular weight of the rubber prior to cross-linking

$\quad R =$ gas constant

$\quad T =$ temperature

and $g =$ 'entanglement factor', i.e. a correction factor to account for any physical restraints that may act as additional 'cross-links' to the actual chemically bonded network

In order to use eqn. 2 to evaluate M_c, it was necessary[6] to determine g by calibrating this equation with a natural rubber vulcanisate of known number of 'chemical' cross-links (i.e. $M_{c(chem)}$). Such a vulcanisate was obtained by using the 'quantitative' cross-linking agent, di-*tert*-butyl peroxide,[6,12] and this is discussed in greater detail in a later section on vulcanisation by organic peroxides (section 3).

Using the techniques described above, it was found possible to determine with some degree of credibility the actual structure of sulphur vulcanisates of natural rubber, both in the presence and absence of accelerators. The significant results are summarised in Tables 2 and 3 for unaccelerated and accelerated systems, respectively. These tables represent experimental data which do not necessarily correspond to commercial practice (unaccelerated systems are not used at all) but illustrate the effect of accelerators and time on the structure of the networks obtained. Thus it can be seen from Table 2 that unaccelerated vulcanisation is very inefficient in the utilisation of sulphur for cross-linking the polymer, the largest proportion of the sulphur being wasted in the formation of cyclic structures on the polymer chains. Furthermore, during the early stages, the cross-links are main poly-sulphidic, but, as the reaction proceeds, the polysulphidic cross-links lose some of their sulphur, which is largely used in creating even more cyclic structures.

A more general picture of the various transformations which the rubber molecules undergo during unaccelerated sulphur vulcanisation is graphi-cally shown in Fig. 1. This illustrates not only the type of structures mentioned in Table 2, i.e. polysulphidic cross-links and cyclic sulphides, but also shows that other events can happen, i.e. formation of vicinal cross-

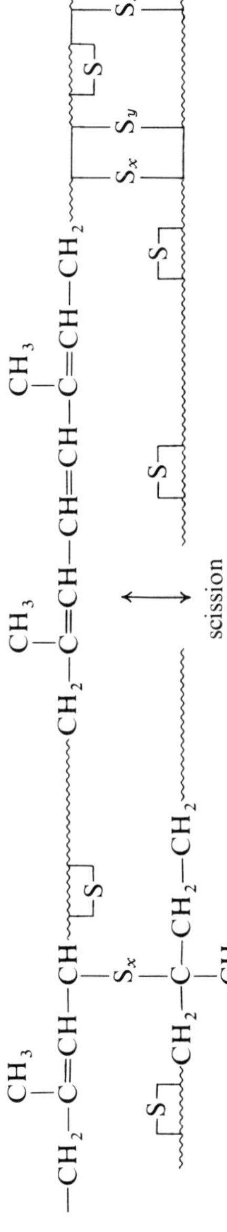

FIG. 1. Main structural features of unaccelerated sulphur vulcanisate of natural rubber.

links (S_x adjacent to S_y), which would act as a single cross-link, chain scission, and dehydrogenation to form conjugated *triene* structures in the chain. In addition, a phenomenon not shown in Fig. 1 is the *isomerisation* of *cis* chain units to *trans*, which is found to happen. This is very important, since it markedly affects chain regularity and thus inhibits the ability of the rubber to crystallise on stretching, which in turn has a profound effect on its mechanical strength.

Before discussing the data of Table 3, for accelerated sulphur vulcanisation, it should be noted that Table 4 represents only one type of rubber compound, and that the results can vary considerably for different compounds and temperatures. Thus Table 4 describes what is known as a very 'efficient' vulcanisation system, in terms of utilisation of sulphur for the formation of cross-links. This is accomplished by using a relatively high level of fatty acid (10 parts instead of the customary 2–3), a somewhat higher level of accelerator, and a relatively low temperature of 100 °C instead of 140 °C. Since this type of recipe results in much too slow a rate, it is quite impractical. However, the practical compounds, with lower amounts of fatty acid and higher temperatures, are less 'efficient' in sulphur utilisation even though they give much faster rates.

A comparison of Tables 2 and 3 shows profound differences between accelerated and unaccelerated vulcanisation. Thus no cyclic sulphide structures are even listed for the accelerated systems (Table 3) and the length of the polysulphidic cross-links is much shorter and decreases rapidly with time. Hence, at the conclusion of the reaction the average cross-link has between 1 and 2 sulphur atoms, i.e. the network contains a mixture of monosulphide and disulphide cross-links. A graphical representation of this type of network is shown in Fig. 2, which is considerably less complicated than Fig. 1, showing a complete absence of cyclic sulphides, polysulphidic cross-links or chain scission. It should be

TABLE 4

VULCANISATION RECIPE FOR TABLE 3

Ingredients	Parts
Rubber	100
Sulphur	1·5
MBT[a]	1·5
Zinc oxide	5·0
Lauric acid	10·0

[a] Mercaptobenzothiazole (an accelerator).

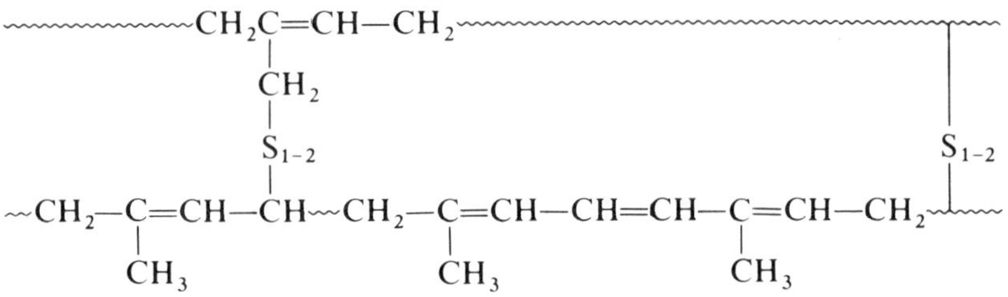

                              M. MORTON

$$\sim\!\!\sim\!\!\sim\!\!CH_2\!\!-\!\!C\!\!=\!\!CH\!\!-\!\!CH_2\!\!\sim\!\!\sim\!\!\sim\!\!\sim\!\!\sim\!\!\sim\!\!\sim\!\!\sim\!\!\sim\!\!\sim$$

with pendant structures:

$$-CH_2-C=CH-CH_2-$$
$$\quad\quad\quad |$$
$$\quad\quad\quad CH_2$$
$$\quad\quad\quad |$$
$$\quad\quad\quad S_{1-2}$$

$$\sim CH_2-C=CH-CH\sim CH_2-C=CH-CH=CH-CH-C=CH-CH_2\sim$$
$$\quad\quad |\quad\quad\quad\quad\quad\quad\quad\quad\quad |\quad\quad\quad\quad\quad\quad\quad\quad\quad |$$
$$\quad\quad CH_3\quad\quad\quad\quad\quad\quad\quad\quad CH_3\quad\quad\quad\quad\quad\quad\quad\quad CH_3$$

FIG. 2. Main structural features of accelerated sulphur vulcanisate of natural rubber.

remembered, of course, that Figs. 1 and 2 are really idealised to emphasise these differences and should not be taken too literally.

In summary then, it can be concluded that the role of the accelerators in the sulphur vulcanisation of natural rubber, besides increasing the reaction rate, is to prevent the excessive side reactions of the sulphur with the rubber, which lead to much modification of the main chains in the form of cyclic sulphide formation, chain scission and double-bond isomerisation. These undesirable reactions not only have an adverse effect on the physical properties of the vulcanisate, but also accelerate the oxidative aging, since the various sulphur structures on the chain act as loci for oxidative attack.

2.1.1. *Mechanism of Accelerated Vulcanisation of Natural Rubber*

On the basis of these experimental results, it was possible to propose a reaction mechanism for the accelerated vulcanisation of natural rubber,[6] and this is shown in eqns. 3–9 as a generalised series of chemical reactions. This mechanism is based on the use of a thiol type of accelerator, such as mercaptobenzothiazole (MBT).

Overall reaction:

$$2RH + S_{x+1} + ZnO \xrightarrow[\text{etc.}]{\text{MBT,}} RS_xR + H_2O + ZnS \tag{3}$$

where RH = rubber

Detailed steps: (XSH = MBT)

$$2XSH + ZnO \longrightarrow XS-Zn-SX + H_2O \tag{4}$$

$$XS-Zn-SX + S_8 \longrightarrow XS_a-Zn-S_bX \tag{5}$$

$$RH + XS_a-Zn-S_bX \longrightarrow RS_aX + ZnS + XS_cH \tag{6}$$

$$2RS_aX \longrightarrow RS_dR + XS_eX \tag{7}$$

$$ZnS + XS_eX \longrightarrow XS_f-Zn-S_gX \quad (e+1=f+g) \tag{8}$$

$$2XS_cH + ZnO \longrightarrow XS_c-Zn-S_cX + H_2O \tag{9}$$

Thus the zinc oxide first forms a zinc salt with the thiol accelerator (eqn. 4), and then reacts with the elemental sulphur to form a zinc polysulphide (eqn. 5). This zinc polysulphide is considered as a 'sulphurating agent' in being instrumental in placing substituent groups on the polymer chain (eqn. 6). This can occur either by substitution, as shown, or by addition to a double bond, although the former seems to predominate, leading to alkenyl sulphide cross-links. This is then followed by reaction 7, which consists of a sulphur–sulphur bond interchange, leading to the formation of cross-links (RS_xR). Equations 8 and 9 describe the regeneration of the sulphurating agent, which can continue the cross-linking reactions (eqns. 6 and 7).

In conclusion, it can be added that step 6, the sulphuration of the rubber, is presumed to occur by a polar (ionic) mechanism, induced by the polar character of the $\overset{\delta+}{Zn}$—$\overset{\delta-}{S}$ bond. The fatty acid is presumably a solubilising agent for the zinc oxide and salts, while amines present in the accelerator or antioxidant could also help this solubilisation by their chelating action.

2.2. Synthetic Rubbers

Much less work of the definitive type described above for sulphur vulcanisation of natural rubber has been carried out on the synthetic rubbers. One comprehensive study of sulphur vulcanisation of styrene–butadiene rubber (SBR) has been reported,[13] using techniques similar to those developed for natural rubber vulcanisates, to establish some detailed knowledge about the network structure. Sulphur vulcanisation of butyl rubber has also been studied to some extent along these lines.[14,15]

In the case of *SBR*, the findings[13] were that the network structures were very similar to those found for natural rubber vulcanisates, i.e. higher polysulphide contents in the early stages of the reaction followed by 'stripping' of the polysulphide cross-links with increasing time of reaction until, at optimum vulcanisation, the cross-links are a mixture of disulphides and monosulphides. Here, too, the unaccelerated systems showed a high consumption of sulphur by side reactions, presumably cyclisation. The MBT-accelerated systems, at their most efficient, seemed to contain fewer sulphur atoms per cross-link than in the case of the analogous natural rubber system, i.e. as low as $0 \cdot 5$ sulphur atom per cross-link. This probably indicates formation of some carbon–carbon cross-links, presumably by a chain addition reaction, leading to cross-links of 'functionality' greater than four, i.e., linking more than two chains.

An interesting sidelight on the effect of different accelerators on the network structure of sulphur vulcanisates is indicated in the above work,[13]

                    M. MORTON

using diphenylguanidine as the accelerator. In the case of both natural rubber and SBR this system leads to cross-links containing a mixture of disulphides and polysulphides, with no evidence of monosulphides or carbon–carbon bonds. Hence these vulcanisates could be completely dissolved in solvents containing an agent capable of cleaving sulphur–sulphur bonds, e.g. di-*n*-butyl phosphite, etc.

In the case of *butyl rubber*, the situation is quite different. Since this is a copolymer of only limited unsaturation, i.e. 1–2% of isoprene copolymerised with isobutene, the reaction with sulphur is much less complex. For example, formation of cyclic sulphides along the chain would be highly improbable since there are very few, if any, vicinal isoprene units. Furthermore, the degree of cross-linking reaches a maximum with time of reaction, presumably when all the available unsaturation is used up, and the vulcanisate need not undergo 'reversion' (degradation) on further heating, as happens in the case of the highly unsaturated rubbers. On this basis, it has been found[14] that, under optimum conditions, sulphur vulcanisates of butyl rubber contain exactly *two* sulphur atoms per cross-link. (The latter is estimated either from sol–gel theory or from a knowledge of the unsaturation in the polymer, both methods agreeing.)

The fact that these cross-links are disulphides was also confirmed by the fact that they could be cleaved by chemical agents known to break sulphur–sulphur bonds. In fact, butyl vulcanisates of this type can be made to undergo reversion at elevated temperatures, and it has been shown[15] that this results in the appearance of thiol groups in the polymer. Such a degraded network can be re-cross-linked to its original state by means of a suitable oxidising agent, e.g. CaO_2, which presumably oxidises any free thiol groups into disulphide cross-linkages.[15]

It should be noted at this point that, because of its limited unsaturation, butyl rubber requires a very active sulphur vulcanisation system, and this is usually accomplished by using larger quantities of accelerators, and also more active accelerators than the usual MBT. The most common of these 'ultra-accelerators' are

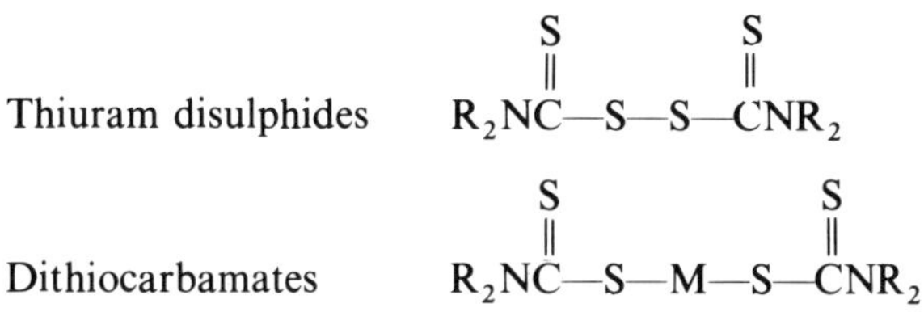

$$\text{Thiuram disulphides} \qquad R_2N\overset{\displaystyle S}{\overset{\|}{C}}\!-\!S\!-\!S\!-\!\overset{\displaystyle S}{\overset{\|}{C}}NR_2$$

$$\text{Dithiocarbamates} \qquad R_2N\overset{\displaystyle S}{\overset{\|}{C}}\!-\!S\!-\!M\!-\!S\!-\!\overset{\displaystyle S}{\overset{\|}{C}}NR_2$$

where M = metal, e.g. zinc or tellurium.

Similar sulphur vulcanisation systems are used for the other well-known elastomer of limited unsaturation, i.e. ethylene–propylene–diene terpolymer (EPDM),[16] although no definitive information is available about the detailed structure of such networks.

Other synthetic rubbers of the polydiene type which are vulcanised by sulphur are, of course, polybutadiene, polyisoprene and nitrile rubber (butadiene–acrylonitrile copolymer). Since no evidence to the contrary is available, one can only assume that the polybutadiene and nitrile form networks of similar structures to those of SBR, while the synthetic polyisoprene behaves like natural rubber.

3. CROSS-LINKING BY PEROXIDES

3.1. Natural Rubber

Although the practical use of organic peroxides to vulcanise natural rubber is quite limited, extensive scientific studies of such cross-linking systems have been carried out because of the relative simplicity of these chemical reactions. The first definitive work on the stoichiometry and mechanism of the reaction was done by Moore et $al.$[10] with di-t-butyl peroxide as the cross-linking agent. Following up on the previous work of Farmer and Moore,[17] they proposed the following series of reactions:

$$(CH_3)_3CO{-}OC(CH_3)_3 \longrightarrow 2(CH_3)_3CO^{\cdot} \tag{10}$$

$$(CH_3)_3CO^{\cdot} + RH \longrightarrow (CH_3)_3COH + R^{\cdot} \tag{11}$$

$$(CH_3)_3CO^{\cdot} \longrightarrow CH_3COCH_3 + CH_3^{\cdot} \tag{12}$$

$$CH_3^{\cdot} + RH \longrightarrow CH_4 + R^{\cdot} \tag{13}$$

$$R^{\cdot} + R^{\cdot} \longrightarrow R{-}R \tag{14}$$

where RH represents a natural rubber molecule. Using model compounds such as 2,6-dimethyl-2,6-octadiene they showed that the hydrocarbon radicals, $R^{\cdot}$ in eqn. 14, invariably couple, rather than disproportionate. In the case of rubber, this would result in a carbon–carbon cross-link.

Assuming, then, that the rubber radicals $R^{\cdot}$ exclusively form cross-links by coupling, as shown, the number of cross-links formed per g of rubber ($\frac{1}{2}M_{c(chem.)}$) is given by $\frac{1}{2}$(mol t-butanol + mol methane) formed per g of rubber. This value was then compared to the value of $\frac{1}{2}M_c$ obtained from physical measurements, i.e. swelling in solvents,[18] or elasticity measurements,[19] in order to obtain a 'calibration curve', such as that shown in

220　　　　　　　　　　　　　　　M. MORTON

Fig. 3, relating the physically measured number of cross-links to the actual
number of chemical cross-links in the network. As expected, it was found
that the former was substantially greater than the latter, especially at low
degrees of cross-linking, presumably due to the chain entanglements which
act as virtual cross-links when 'trapped' in the network. Of course, as more
chemical cross-links are introduced into the network, the contribution of
the entanglements becomes less important.

Thus it appears that di-*t*-butyl peroxide is a 'quantitative' cross-linking
agent, *one* mol of the peroxide accounting for *one* cross-link, in accordance
with the scheme shown in eqns. 10–14. Similar stoichiometry was

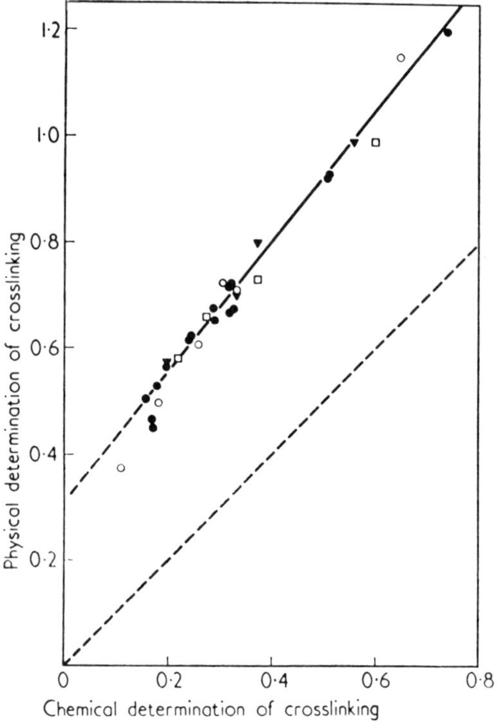

FIG. 3.　Comparison of physical ($\frac{1}{2}M_c \times 10^4$) and chemical ($\frac{1}{2}M_{c(chem.)} \times 10^4$)
values of degree of cross-linking of vulcanisates obtained from di-*t*-butyl peroxide
natural rubber system. ——, Experimental; – – –, theoretical. Vulcanisation of ●,
140°C; ○, 130°C; □, 120°C; ▼, 110°C. Reproduced from reference 6 with
permission.

demonstrated[20] for the more common dicumyl peroxide (Dicup) used commercially. The only complication that the above chemical equations do not take into account is the possibility of chain scission during the vulcanisation, and this would vitiate any determination of cross-link density by physical measurements. However, sol–gel measurements on natural rubber vulcanisates cross-linked by dicumyl peroxide showed[21,22] that this is a very minor factor.

On the basis of the above experimental results, it is possible to deduce the structure of peroxide cross-linked natural rubber vulcanisates, and this is shown schematically in Fig. 4. Thus this network consists primarily of di-alkenyl carbon–carbon cross-links, but can also contain some cyclic structures, presumably formed by *intra-molecular* addition of the alkenyl radicals to double bonds. It should be noted that such cyclisation reactions do not alter the stoichiometry of the radical coupling reaction (eqn. 14). Little is known about the frequency of these cyclisation reactions, but they must be a minor occurrence, since the basic structure of the *cis*-1,4-polyisoprene chains is not altered to any extent, as shown by the crystallisation behaviour of the vulcanised rubber, which is very similar to that of the unvulcanised rubber. This is in sharp contrast to the sulphur vulcanisates, which are markedly different from the original rubber because of the many side reactions, including cyclisation, isomerisation, etc.

3.2. Synthetic Rubbers
3.2.1. Saturated Elastomers

Other than synthetic *cis*-1,4-polyisoprene, which presumably reacts with peroxides in a similar manner to natural rubber, the synthetic rubbers undergo various reactions, depending on their chemical structure, during cross-linking by peroxides. As indicated in Table 1, peroxides may be used to cross-link either saturated or unsaturated elastomers, but, in practice, they are used primarily for the saturated variety, i.e. silicones, urethanes and highly filled polyethylene. This is because the unsaturated elastomers are more conveniently vulcanised with sulphur. Actually, the silicones and urethanes are also available in grades which contain a modest amount of unsaturation, and can therefore be vulcanised with sulphur. The sole exception is butyl rubber, which *cannot* be cross-linked by peroxides since it undergoes serious degradation when so treated.

This is unlike the case of the ethylene–propylene rubbers, which, like butyl, contain a limited amount of unsaturation, but can be vulcanised either by *sulphur or peroxides*.

The classic example of the cross-linking of a saturated polymer by

M. MORTON

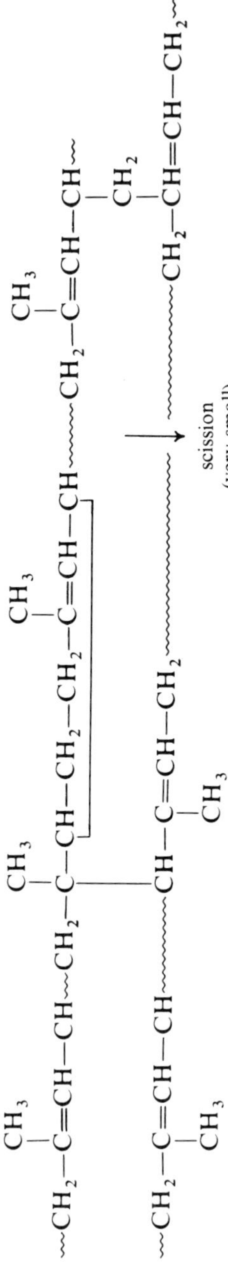

Fig. 4. Structure of peroxide cross-linked natural rubber.

peroxides is provided by the silicones, where peroxide cross-linking is pictured as follows:

$$ROOR \longrightarrow 2RO\cdot \tag{15}$$

$$RO\cdot + \text{$\sim\!\!$Si}\!-\!\text{O}\!\sim \longrightarrow ROH + \text{$\sim\!\!$Si}\!-\!\text{O}\!\sim \tag{16}$$

with the left silicon bearing CH_3 groups above and below, and the right silicon bearing CH_3 above and $CH_2\cdot$ below.

$$2\,\text{$\sim\!\!$Si}\!-\!\text{O}\!\sim \longrightarrow \text{$\sim\!\!$Si}\!-\!\text{O}\!\sim \tag{17}$$

with the reactant silicon bearing CH_3 above and $CH_2\cdot$ below, and the product silicon bearing CH_3 above and a $CH_2\!-\!CH_2$ bridge below connecting to $\sim\!\!$Si$-$O$\sim$.

This proposed mechanism is not necessarily 100% efficient. Thus the dialkyl or dialkaryl peroxides, such as di-t-butyl or dicumyl peroxides, are very inefficient, leading to very poor vulcanisation of the saturated silicones, presumably because of their lack of reactivity in abstracting hydrogen from the methyl groups. On the other hand, the diaroyl peroxides (e.g. benzoyl) are about 50% efficient[23] in terms of reaction 16 above and about 40% efficient in reaction 17. Thus only about 0·1–0·3 cross-links are obtained for each peroxide decomposed.[24] The low efficiencies described above, when compared with the 100% efficiency found for the alkyl and alkaryl peroxides in natural rubber, are undoubtedly due to the high energies involved in hydrogen abstraction from alkyl compared with alkenyl groups.

3.2.2. Unsaturated Elastomers

Peroxides are not used extensively in the vulcanisation of unsaturated synthetic rubbers because of the superiority of sulphur vulcanisation. However, the peroxide vulcanisation of polybutadiene and butadiene-based rubbers has been the subject of considerable interest and study.[25–39] This widespread interest was undoubtedly due to the fact that the butadiene unit in polymers was not found to cross-link by the simple mechanism outlined for polyisoprene in eqns. 10–14. Although polybutadiene contains allylic hydrogen, just like polyisoprene, it also has a much more reactive double bond, unobstructed by the substituent methyl group present in polyisoprene. Furthermore, polybutadiene and its copolymers may have varying amounts of 1,2-units, unlike the 100% *cis*-1,4-units in natural rubber, and it is intuitively obvious, and experimentally confirmed, that the substituent vinyl groups would be considerably more reactive than the internal double bonds.

In view of the above, a different mechanism could be proposed for the cross-linking of butadiene units, based on an *addition reaction* similar to the propagation step in vinyl polymerisation, and this could operate in *competition* with the allylic hydrogen abstraction mechanism which apparently occurs exclusively with natural rubber. The various steps in the 'polymerisation' type of cross-linking of polybutadiene are shown in eqns. 18–22.

Initiation

$$ROOR \longrightarrow 2RO^{\cdot} \tag{18}$$

$$RO^{\cdot} + \begin{array}{c} \sim\!\!\!CH_2\!-\!CH\!\sim \\ | \\ CH_2\!\!=\!\!CH \end{array} \longrightarrow \begin{array}{c} RO\!-\!CH_2\!-\!CH^{\cdot} \\ | \\ \sim\!\!\!CH_2\!-\!CH\!\sim \end{array} \tag{19}$$

Propagation

$$\begin{array}{c} RO\!-\!CH_2\!-\!CH^{\cdot} \\ | \\ \sim\!\!\!CH_2\!-\!CH\!\sim \end{array} + \begin{array}{c} CH_2\!\!=\!\!CH \\ | \\ \sim\!\!\!CH_2\!-\!CH\!\sim \end{array} \rightarrow \cdots \rightarrow RO\!\!\left[\begin{array}{c} CH_2\!-\!CH \\ | \\ CH_2\!-\!CH \end{array}\right]\!\!\begin{array}{c} CH_2\!-\!CH^{\cdot} \\ | \\ CH_2\!-\!CH\!\sim \end{array}_x \tag{20}$$

Transfer

$$\begin{array}{c} RO\!\!\left[\begin{array}{c} CH_2\!-\!CH \\ | \\ CH_2\!-\!CH \end{array}\right]_x\!\!\begin{array}{c} CH_2\!-\!CH^{\cdot} \\ | \\ CH_2\!-\!CH\!\sim \end{array} + \sim\!\!\!CH_2\!-\!CH\!\!=\!\!CH\!-\!CH_2\!\sim \\[2ex] RO\!\!\left[\begin{array}{c} CH_2\!-\!CH \\ | \\ CH_2\!-\!CH \end{array}\right]_x\!\!\begin{array}{c} CH_2\!-\!CH_2 \\ | \\ CH_2\!-\!CH\!\sim \end{array} + \sim\!\!\!CH_2\!-\!CH\!\!=\!\!CH\!-\!\dot{C}H\!\sim \end{array} \tag{21}$$

Termination

By combination or disproportionation of radicals $\qquad$ (22)

The addition reactions 19 and 20 are here shown to occur exclusively with the vinyl double bonds (1,2-units) in the polymer, since these would be assumed to be much more reactive than their 1,4-counterparts, but reactions 19 and 20 may not be exclusive. These reactions, then, compete with the 'transfer' reaction (21), which represents the hydrogen *abstraction* reaction (eqn. 11) previously indicated for natural rubber. This can, of course, also occur with the initiating radicals $RO^{\cdot}$ in the case of polybutadiene, but it is not shown, for the sake of simplicity. It should be noted that eqn. 21 shows abstraction of hydrogen from a 1,4-chain unit, but it obviously can occur with a 1,2-unit as well, even though there are four allylic hydrogens in each 1,4-unit and only two such hydrogens in every 1,2-unit.

Thus it is apparent that a mechanism operating in accordance with eqns. 18–22 would lead to the following situation: (a) the cross-linking of *more than two* chains at each junction point, by reaction 20, i.e. 'polyfunctional' cross-links, instead of the *'tetra*functional' network junctions found in natural rubber (*four* chains emanating from each juncture), and (b) a high 'efficiency' of cross-linking, i.e. a larger number of cross-links from each peroxide molecule, regardless of the relative frequency of occurrence of reactions 20 and 21. The latter feature had already been deduced, at least qualitatively, by the observations that peroxides are much more efficient in cross-linking polybutadiene than natural rubber, in terms of the amount of peroxide needed to attain a given cross-link density.[25,26,28,30–33]

Most of the conclusions described above were reached by determining the degree of cross-linking by tension or swelling measurements, and these methods may be questionable when applied to networks having cross-links of high functionality. In order to avoid this problem, Scott[34] developed two alternate methods for measuring the cross-linking kinetics, one based on sol–gel theory (post-gelation), and the other on molecular weight changes before the gel point is reached (pre-gelation).

Post-gelation treatment.[34] The relation between sol content and degree of cross-linking of a polymer of *monodisperse* molecular weight was expressed as follows:

$$yI/i = yq = -\ln w_{\mathrm{s}} \left/ \left[I - \sum_{1}^{\infty} w_x w_{\mathrm{s}}^{x-1} \right] \right. \tag{23}$$

where y = number of units in the primary chain (monodisperse)
$\quad\quad I$ = number of free radicals consumed, expressed as mol of radicals consumed per mol of chain units
$\quad\quad i$ = probability of termination of the cross-linking reaction
$\quad\quad q$ = fraction of cross-linked units in the vulcanisate
$\quad\quad w_{\mathrm{s}}$ = wt. fraction of sol
$\quad\quad w_x$ = wt. fraction of cross-linked units occurring in cross-links composed of x cross-linked units

Equation 23 can be used to determine the value of $1/i$ (kinetic chain length (number of chain units cross-linked per radical)) by using a monodisperse polymer and determining the weight fraction of sol, w_{s}, at different extents of cross-linking, I, assuming 100% efficiency for the initiation step (eqn. 19).

A study of this type was carried out on a series of polybutadienes prepared by organolithium polymerisation[38] to have varying vinyl contents, and the values of kinetic chain length, $1/i$, obtained are shown in Table 5. Since the kinetic chain length represents the number of chains cross-linked by each radical, it is clearly evident from these data that it is the vinyl groups in polybutadiene that are the more reactive species. However, the fact that even the 99% 1,4-polybutadiene still shows *more than two* chains cross-linked per free radical indicates that the 1,4-units are reactive enough to be attacked by addition of such radicals. It should be noted, however, that the kinetic chain length values in Table 5 do not necessarily represent the number of chains linked together by *each* cross-link, since there is no information about the extent of the transfer reaction (eqn. 21). Such information can be obtained by using the 'pre-gelation' treatment developed by Scott.[34]

Pre-gelation treatment.[34] It is obvious that, prior to the onset of gel, a polymer undergoing cross-linking will experience a continuous change in molecular weight. The relation derived for this change was expressed as follows, for polymers of *monodisperse* molecular weight:

$$1/Y'_n = 1/Y_n - (X_n - 1)I/X_n i = 1/Y_n - R_1 I \qquad (24)$$

$$1/Y'_w = 1/Y_w - (X_w - 1)I/i = 1/Y_w - R_2 I \qquad (25)$$

where Y_n and Y_w are, respectively, the number-average and weight-average number of units per original chains
Y'_n and Y'_w are the corresponding values *after* cross-linking
X_n and X_w are, respectively, the number-average and weight-average number of chains per cross-link, i.e. the cross-link 'functionality'.
I and i have the same connotation as in eqn. 23

Thus from a knowledge of Y_n, Y_w, Y'_n and Y'_w, as well as I, it is possible to solve the simultaneous equations where i and X_n (or X_w) are the unknowns.

As in the case of the post-gelation experiments, a series of monodisperse polybutadienes of varying vinyl content, prepared by organolithium polymerisation, were treated with dicumyl peroxide for varying periods of time, and the number-average and weight-average molecular weights determined by osmometry and light scattering, respectively.[39] The results

TABLE 5
CROSS-LINKING OF POLYBUTADIENES BY DICUMYL PEROXIDE[38] (110 °C)

Vinyl content in polymer (%)	Kinetic chain length (1/i)
1[a]	2·9
9·5	3·1
30	10·3
76	19·0
88	23·8

[a] Fraction from a commercial high-*cis* polymer.

are shown in Table 6. It can be seen that here, too, the kinetic chain length increases with increasing vinyl content of the polymer, as would be expected in a 'polymerisation' type of cross-linking reaction. The values are fairly similar to those shown in Table 5, although they seem to rise to a much higher value for the 88 % vinyl polymer. This may have been due to experimental difficulties in determining the molecular weight by light scattering when it is changing rapidly, and possibly forming a microgel.

In any event, the relatively low values of the cross-link functionality indicate strongly that a chain-transfer reaction of the type shown in eqn. 21 predominates over termination. Hence, even though the efficiency of cross-linking may be high (high values of kinetic chain length), only a few chains (3 or 4) are cross-linking together at a single juncture point. In other words, a peroxide molecule may be responsible for a large number of cross-links

TABLE 6
CROSS-LINKING OF POLYBUTADIENE BY DICUMYL PEROXIDE[39] (110 °C)

Vinyl content in polymer (%)	Kinetic chain length	Cross-link functionality	
		X_n	X_w
8	2·1	1·9	2·7
21	3·1	2·0	2·9
35	4·3	2·1	3·2
45	8·2	2·3	3·5
58	8·3	3·0	5·0
88	43	3·8	6·7

(20 or 30), but the functionality of each cross-link is only a little greater than the classical tetrafunctional cross-link found in natural rubber vulcanisates.

4. CROSS-LINKING BY RADIATION

The radiation chemistry that leads to cross-linking polymers has been extensively studied and widely published. It is quite complex and depends greatly on the type of radiation and structure of the polymer, with many side reactions, e.g. chain scission, isomerisation, dehydrogenation, etc. A good review is available of the specific application of radiation to the cross-linking of elastomers.[40] Because of the broad range of this field and the similarity of the cross-links introduced by radiation to those formed by peroxides, i.e. the carbon–carbon single bond, this topic will not be dealt with here.

5. CROSS-LINKING BY METAL OXIDES

As indicated in Table 1, metal oxides (e.g. zinc oxide, magnesium oxide, etc.) are used to cross-link two types of polymers: (a) those containing halogen atoms, and (b) those containing carboxyl groups.

The most important instance of cross-linking by metal oxides involves polychloroprene (neoprene),[41] where the metal oxide (ZnO or MgO) reacts with allylic chlorine of the small percentage of 1,2-units in the chain, thus:

$$
\begin{array}{c}
CH{=}CH_2 \\
| \\
{\sim}CH_2{-}C{\sim} \\
| \\
Cl + ZnO \longrightarrow \\
\\
Cl \\
| \\
{\sim}CH_2{-}C{\sim} \\
| \\
CH{=}CH_2
\end{array}
\qquad
\begin{array}{c}
CH{=}CH_2 \\
| \\
{\sim}CH_2{-}C{\sim} \\
| \\
O \qquad + ZnCl_2 \qquad (26) \\
| \\
{\sim}CH_2{-}C{\sim} \\
| \\
CH{=}CH_2
\end{array}
$$

Another chlorinated elastomer which is cross-linked by metal oxides is chlorosulphonated polyethylene (Hypalon). The small amount of sulphonyl chloride groups are considered as the reactive species in forming metal sulphonate cross-links,[42] thus:

$$\begin{array}{ccc}
\text{\sim\sim CH}_2\text{—CH\sim\sim} & & \text{\sim\sim CH}_2\text{—CH\sim\sim} \\
| & & | \\
\text{SO}_2 & & \text{SO}_2 \\
| & & | \\
\text{Cl} + 2\text{ZnO} \longrightarrow & & \text{O} \quad + \text{ZnCl}_2 \\
| & & | \\
\text{Cl} & & \text{Zn} \\
| & & | \\
\text{SO}_2 & & \text{O} \\
| & & | \\
\text{\sim\sim CH}_2\text{—CH\sim\sim} & & \text{SO}_2 \\
& & | \\
& & \text{\sim\sim CH}_2\text{—CH\sim\sim}
\end{array} \qquad (27)$$

Chlorbutyl, which is obtained by chlorination of isoprene units in butyl rubber, can also be cross-linked by metal oxides. Here the chlorine is present as allylic chlorine, in several isomeric forms, undoubtedly formed by elimination of HCl. It has been suggested[43] that the cross-linking is actually caused by zinc chloride formed from zinc oxide, and that this leads to carbon–carbon cross-links by a cationic mechanism, thus:

$$\begin{array}{l}
\underset{\qquad\qquad |}{\overset{\text{CH}_3}{}}\\
\text{\sim\sim CH}{=}\underset{|}{\text{C}}\text{—}\underset{|}{\text{CH}}\text{—CH}_2\text{\sim\sim} + \text{ZnCl}_2 \longrightarrow \text{\sim\sim CH}{=}\overset{\text{CH}_3}{\underset{}{\text{C}}}\text{—}\overset{+}{\text{CH}}\text{—CH}_2\text{\sim\sim} \\
\qquad\qquad\;\; \text{Cl} \qquad\qquad\qquad\qquad\qquad\qquad\qquad\qquad (\text{ZnCl}_3)^{-}
\end{array}$$

$$\text{\sim\sim CH}{=}\overset{\text{CH}_3}{\text{C}}\text{—}\overset{+}{\text{CH}}\text{—CH}_2\text{\sim\sim} + \text{\sim\sim CH}{=}\overset{\text{CH}_3}{\text{C}}\text{—}\underset{\underset{\text{Cl}}{|}}{\text{CH}}\text{—CH}_2\text{\sim\sim} \longrightarrow$$
$$(\text{ZnCl}_3)^{-}$$

$$\begin{array}{ll}
\overset{\text{CH}_3}{} & \overset{\text{CH}_3}{} \\
\text{\sim\sim CH}{=}\text{C—CH—CH}_2\text{\sim\sim} & \text{\sim\sim CH}{=}\text{C—CH—CH}_2\text{\sim\sim} \\
\qquad\quad | & \qquad\quad | \\
\qquad\quad | \quad \overset{\text{CH}_3}{} & \longrightarrow \qquad\quad | \quad \overset{\text{CH}_3}{} \\
\text{\sim\sim CH—}\overset{+}{\text{C}}\text{—CH—CH}_2\text{\sim\sim} & \text{\sim\sim C}{=}\text{C—CH—CH}_2\text{\sim\sim} \\
(\text{ZnCl}_3)^{-} \; | & \qquad\qquad\; | \\
\qquad\qquad \text{Cl} & \qquad\qquad \text{Cl}
\end{array}$$
$$+ \text{HCl} + \text{ZnCl}_2$$

$$(28)$$

Metal oxides are also used to cross-link carboxyl-containing elastomers, such as copolymers of butadiene and methacrylic acid. Metal carboxylate salt bridges are formed,[44] thus:

$$\begin{array}{c}
\text{\textasciitilde}CH_2-CH\text{\textasciitilde} \\
| \\
COOH \\
\\
+ \; ZnO \; \longrightarrow \\
\\
COOH \\
| \\
\text{\textasciitilde}CH_2-CH\text{\textasciitilde}
\end{array}
\qquad
\begin{array}{c}
\text{\textasciitilde}CH_2-CH\text{\textasciitilde} \\
| \\
C{=}O \\
| \\
O \qquad + \; H_2O \qquad (29)\\
| \\
Zn \\
| \\
O \\
| \\
C{=}O \\
| \\
\text{\textasciitilde}CH_2-CH\text{\textasciitilde}
\end{array}$$

Such salt-type cross-links are considered to be relatively labile and can undergo considerable bond interchange.[44]

6.   CROSS-LINKING BY DI- (OR POLY-) AMINES

Halogen-containing polymers can also be cross-linked by diamines (or polyamines). Among the more important elastomers that can be cross-linked by these agents are polychloroprene, chlorbutyl and fluoro-rubbers.

In the case of chlorbutyl, the reaction is relatively straightforward, between diamines and the secondary allyl chloride of the chlorbutyl, thus:

$$\begin{array}{c}
\quad\quad CH_3 \\
\quad\quad | \\
\text{\textasciitilde}CH{=}C-CH-CH_2\text{\textasciitilde} \\
\quad\quad\quad\quad | \\
\quad\quad\quad\quad Cl \\
\\
\quad + \; H_2NRNH_2 \; \longrightarrow \\
\\
\quad\quad Cl \\
\quad\quad | \\
\text{\textasciitilde}CH{=}C-CH-CH_2\text{\textasciitilde} \\
\quad\quad | \\
\quad\quad CH_3
\end{array}
\qquad
\begin{array}{c}
\quad\quad CH_3 \\
\quad\quad | \\
\text{\textasciitilde}CH{=}C-CH-CH_2\text{\textasciitilde} \\
\quad\quad\quad\quad | \\
\quad\quad\quad\quad NH \\
\quad\quad\quad\quad | \\
\quad\quad\quad\quad R \qquad + \; 2HCl \\
\quad\quad\quad\quad | \\
\quad\quad\quad\quad NH \\
\quad\quad\quad\quad | \\
\text{\textasciitilde}CH{=}C-CH-CH_2\text{\textasciitilde} \\
\quad\quad | \\
\quad\quad CH_3 \qquad\qquad (30)
\end{array}$$

The above reaction (30) is presumed to go through an intermediate amine hydrochloride formation. An HCl scavenger is generally used as well.

Polychloroprene can also be vulcanised efficiently with diamines which react with the allylic chlorine,[45] but the reaction involves some isomerisation of the double bond, as follows:

$$
\begin{array}{c}
\overset{\displaystyle \overset{\S}{CH_2}}{\underset{\S}{2Cl-C-CH=CH_2}} + H_2N-R-NH_2 \\
\downarrow \\
\overset{\displaystyle \overset{\S}{CH_2}}{\underset{\S}{C=CH-CH_2-NH-R-NH-CH_2-CH=C}} + 2HCl
\end{array}
\qquad (31)
$$

Zinc oxide is often used as acid acceptor.

The cross-linking of fluororubbers is also carried out by diamines, but the mechanism is much more complex. Apparently the cross-linking occurs in stages[46] as the temperature is raised from 121 °C (250 °F) in the mould to the final 204 °C (400 °F) in an oven 'post-cure'. The first stage leads to formation of diamine cross-links, with elimination of hydrogen fluoride, as shown in eqn. 32:

$$
\begin{array}{c}
\sim\!CF_2-CH_2-\overset{\displaystyle CF_3}{\overset{|}{CF}}-CF_2\!\sim \\[2ex]
+\,H_2N-R-NH_2 \longrightarrow \\[2ex]
\sim\!CF_2-CH_2-\underset{\displaystyle CF_3}{\underset{|}{CF}}-CF_2\!\sim
\end{array}
$$

$$
\begin{array}{c}
\sim\!CF_2-\overset{\displaystyle CF_3}{\overset{}{CH}}-CH-CF_2\!\sim \\
| \\
NH \\
| \\
R \qquad\qquad +\,2HF \\
| \\
NH \\
| \\
\sim\!CF_2-CH-CH-CF_2\!\sim \\
| \\
CF_3
\end{array}
\qquad (32)
$$

At higher temperatures, in the second stage, more hydrogen fluoride is eliminated, with formation of *conjugated* double bonds, which undergo a Diels–Alder addition to form more cross-links, and these are eventually converted to aromatic cross-links. Hence it is the final oven treatment at 204 °C (400 °F) that results in the heat-stable cross-links required for these high temperature polymers.

7.   OTHER METHODS OF CROSS-LINKING

Elastomers based on dienes can be cross-linked by the above type of organic reagents, provided the polymer chain has *allylic* hydrogen atoms. The reaction mechanism is similar in all cases and involves an 'ene' reaction[47,48] whereby *substitution* occurs at the α-*carbon* adjacent to the double bond,[49] resulting in a carbon–carbon cross-link. Such cross-links are generally more heat-stable than those involving carbon–sulphur bonds.

7.1. Phenolic Resins

The cross-linking agent here has the general structure

$$\left[\; X\!-\!CH_2\!-\!\underset{R}{\overset{O}{\bigcirc}}\!-\!CH_2X \; \right]_n$$

where X is OH or halogen, R is an alkyl group, and n is small, depending on the resin used. The reaction is shown in eqn. 33.

$$2 \;\underset{CH_2}{\overset{-C}{\underset{|}{\overset{\parallel}{-C}}}}\; + \; X\left[CH_2\!-\!\underset{R}{\bigcirc}\!-\!CH_2X\right]_n \;\rightarrow\; -\underset{CH}{\overset{-C}{\underset{\parallel}{\underset{-C}{|}}}}\left[CH_2\!-\!\underset{R}{\overset{OH}{\bigcirc}}\!-\!CH_2\right]_n\!-\!\underset{CH}{\overset{C-}{\underset{\parallel}{C-}}}$$

$$\tag{33}$$

It should be noted that this type of reaction results in a *shifting* of the double bond in the polymer chain, a characteristic of 'ene' reactions. This type of cross-linking is utilised especially for butyl rubber (and other low unsaturation polymers) since the network is then much more resistant to degradation at high temperatures, compared with sulphur vulcanisates.

7.2. Quinone Derivatives

A similar reaction occurs with the use of quinone derivatives, principally *quinone dioxime*; the reaction is shown in eqn. 34. This presupposes the prior oxidation of the dioxime to a dinitroso compound,[48] and this usually requires some peroxide or other oxidising agent. Dinitrosobenzene alone is very effective in cross-linking diene elastomers. The conversion of the dioxime cross-link to an amine has not been elucidated as yet.[48]

$$(34)$$

7.3. Maleimide Derivatives

Various bis-maleimides also cross-link unsaturated elastomers. Peroxides are frequently used to initiate the reaction which then proceeds by a *free-radical* mechanism[50] as follows: the initiator radical abstracts hydrogen from the α-carbon, the resulting free radical adds to the carbon–carbon double bond of one of the maleimide groups, which then abstracts hydrogen from another α-carbon of the polymer chain, and the same thing happens with the other maleimide group. In this mechanism there is no *isomerisation* of the polymer double bonds.

However, at sufficiently high temperatures, the bis-maleimides will cross-link unsaturated polymers without the aid of free radicals, and presumably operate by the 'ene' mechanism, leading to a *shifting* of the double bond of the polymer, as in eqn. 35.

In conclusion, it should be mentioned that there are other variations of the cross-linking methods described in this section, and these have been reviewed recently.[48]

$$
\begin{array}{ccc}
\sim C{=}C{-}CH_2\sim & \begin{array}{c} HC{=}CH \\ CO \quad CO \\ N \\ R \end{array} & \\
+ & N & \longrightarrow \\
\sim C{=}C{-}CH_2\sim & \begin{array}{c} CO \quad CO \\ HC{=}CH \end{array} &
\end{array} \tag{35}
$$

8. EFFECT OF NATURE OF CROSS-LINK ON STRENGTH OF ELASTOMERS

Since the strength of a rubber vulcanisate (tensile, tear, abrasion) is of utmost concern, there has been much speculation about the ability of the type of chemical bond comprising the 'cross-link' to affect this property, i.e. 'strong' versus 'weak' cross-links. However, this question has been very difficult to resolve, since the chemical reactions involved in introducing the different cross-links are quite different and may cause other changes in the network as well, e.g. chain scission, chain modification, etc. Hence it has been virtually impossible to relate differences in strength of rubber vulcanisates *specifically* to differences in the nature of the cross-link.

One instance, however, which seemed to indicate such a direct relation was the case of natural rubber. It was noted some time ago[51] that sulphur vulcanisates of natural rubber have substantially higher strength (at optimum vulcanisation) than does peroxide- (or radiation-) cross-linked material. This was ascribed to the 'lability' of the sulphur–sulphur bond, which can presumably rupture and reform, by bond interchange, thus absorbing the stress and preventing failure, a possibility which is presumably not available to the carbon–carbon bond of peroxide vulcanisates. A similar hypothesis was proposed[52] for the differences in flexing failure between sulphur and peroxide vulcanisates of SBR.

This rather attractive theory about the strength advantages of 'labile' cross-links has, however, been largely disproven by the results of two convincing investigations.[53,54] In the first, Tobolsky and Lyons[53] showed, by means of stress relaxation studies, that there was no evidence that the sulphur–sulphur bonds present in a sulphur vulcanisate were capable of yielding under *mechanical stress* any more than the carbon–carbon cross-links in peroxide vulcanisates, although such sulphur–sulphur bonds are known to undergo bond interchange at *elevated temperatures*. They proposed, instead, that this thermal lability of sulphur–sulphur cross-links made it possible for the network, during its formative stage, to rearrange so as to reduce any built-in stresses, a process not available in peroxide cross-linking. Hence the latter leads to networks with built-in stresses which can lead to premature failure.

The other definitive study was by Lal,[54] who prepared both sulphur and peroxide vulcanisates of natural rubber and noted the effect of the presence of either sulphur–sulphur bonds, carbon–carbon bonds, or a mixture of the two. In the first place, it was noted that extraction of sulphur vulcanisates by triphenylphosphine, which is known to 'strip' sulphur from polysulphides, did not reduce the strength of the network. Furthermore, the introduction of sulphur–sulphur cross-links into a peroxide vulcanisate, by means of a 'second-stage' vulcanisation, did *not* enhance the strength, as might have been expected. Hence there does not appear to be any support for the theory that 'weak' cross-links actually lead to strong rubber networks, and the reasons for such differences in strength must lie elsewhere.

REFERENCES

1. ARMSTRONG, R. T., LITTLE, J. R. and DOAK, K. W., *Ind. Eng. Chem.*, **36** (1944), p. 628; *Rubber Chem. Technol.*, **17** (1944), p. 788.
2. SELKER, M. L. and KEMP, A. R., *Ind. Eng. Chem.*, **36** (1944), p. 16, *ibid.*, **39** (1947), p. 985.
3. FARMER, E. H. and SHIPLEY, F. W., *J. Chem. Soc.* (1947), p. 1519.
4. FARMER, E. H., *J. Soc. Chem. Ind.*, **66** (1947), p. 36.
5. SHELTON, J. R. and McDONEL, E. T., *Rubber Chem. Technol.*, **33** (1960), p. 342.
6. BATEMAN, L., MOORE, C. G., PORTER, M. and SAVILLE, B., in *The Chemistry and Physics of Rubberlike Substances*, Bateman, L. (Ed.), London, MacLaren Press, 1963, Chapter 15, p. 449.
7. BATEMAN, L., GLAZEBROOK, R. W., MOORE, C. G., PORTER, M., ROSS, G. W. and SAVILLE, B., *J. Chem. Soc.* (1958), pp. 2838, 2846; *Rubber Chem. Technol.*, **31** (1958), p. 1055.

8. BATEMAN, L., MOORE, C. G. and PORTER, M., *J. Chem. Soc.* (1958), p. 2866; *Rubber Chem. Technol.*, **31** (1958), p. 1090.

9. BATEMAN, L., GLAZEBROOK, R. W. and MOORE, C. G., *J. Appl. Polym. Sci.*, **1** (1959), p. 257; *Rubber Chem. Technol.*, **31** (1958), p. 1065.

10. MOORE, C. G., MULLINS, L. and SWIFT, P. McL., *J. Appl. Polym. Sci.*, **5** (1961), p. 293.

11. SAVILLE, B. and WATSON, A. A., *Rubber Revs.*, **40** (1967), p. 100.

12. MOORE, C. G. and WATSON, W. F., *J. Polym. Sci.*, **19** (1956), p. 237.

13. GAN, L. M., Ph.D. Dissertation, University of Akron, 1967.

14. ZAPP, R. L., DECKER, R. A., DYROFF, M. S. and RAYNER, H. A., *J. Polym. Sci.*, **6** (1951), p. 331.

15. ZAPP, R. L. and FORD, F. P., *J. Polym. Sci.*, **9** (1952), p. 97.

16. BORG, E. L., in *Rubber Technology*, Morton, M. (Ed.), New York, Van Nostrand Reinhold, 1973, p. 220.

17. FARMER, E. H. and MOORE, C. G., *J. Chem. Soc.* (1951), pp. 131, 142.

18. MULLINS, L., *J. Polym. Sci.*, **19** (1956), p. 225.

19. MULLINS, L., *J. Appl. Polym. Sci.*, **2** (1959), p. 1.

20. PARKS, C. R. and LORENZ, O., *J. Polym. Sci.*, **50** (1961), pp. 287, 299.

21. SCOTT, K. W., *J. Polym. Sci.*, **58** (1962), p. 517.

22. BRISTOW, G. M., *J. Appl. Polym. Sci.*, **7** (1963), p. 1023.

23. LEWIS, F. M., in *Polymer Chemistry of Synthetic Elastomers*, Kennedy, J. P. and Tornqvist, E. G. M. (Eds.), New York, John Wiley, 1969, Chapter 8.

24. BOBEAR, W. J., *Rubber Chem. Technol.*, **40** (1967), p. 1560; also in *Rubber Technology*, Morton, M. (Ed.), New York, Van Nostrand Reinhold, 1973, p. 372.

25. VAUGHAN, G., EAVES, D. E. and COOPER, W., *Polymer*, **2** (1961), p. 235.

26. COOPER, W., EAVES, D. E. and VAUGHAN, G., *J. Polym. Sci.*, **59** (1962), p. 241.

27. SCHEELE, W., SCHULZE, W. and HILLMER, K.-H., *Kautschuk Gummi*, **15** (1962), p. WT57.

28. VAN DER HOFF, B. M. E., *Ind. Eng. Chem. Prod. Res. Develop.*, **2** (1963), p. 273.

29. KRAUS, G., *J. Appl. Polym. Sci.*, **7** (1963), p. 1257.

30. LOAN, L. D., *J. Appl. Polym. Sci.*, **7** (1963), p. 2259.

31. LOAN, L. D. and SCANLAN, J., *Rubber Plastics Age*, **44** (1963), p. 1315.

32. HUMMEL, K. and KAISER, G., *Kautschuk Gummi*, **16** (1963), p. 436; *Rubber Chem. Technol.*, **38** (1965), p. 581.

33. VAN DER HOFF, B. M. E., *Appl. Polym. Symp.*, **7** (1968), p. 21.

34. SCOTT, K. W., *Polym. Preprints*, **5** (1964), p. 756.

35. BHATNAGAR, S. K. and BANERJEE, S., *Makromol. Chem.*, **109** (1967), p. 217.

36. MANIK, S. P. and BANERJEE, S., *Angew. Makromol. Chem.*, **6** (1969), p. 171.

37. HERGENROTHER, W. L., *Polym. Preprints*, **11** (1970), p. 834.

38. WEST, J. C., Ph.D. Dissertation, University of Akron, 1971; see also MORTON, M. and WEST, J. C., *100th Meeting, Rubber Division, Amer. Chem. Soc., Cleveland*, 1971, Paper No. 44.

39. MALOTKY, L. O., Ph.D. Dissertation, University of Akron, 1973; see also MALOTKY, L. O. and MORTON, M., *Polym. Preprints*, **15** (1974), p. 714.

40. TURNER, D. J., in *The Chemistry and Physics of Rubberlike Substances*, Bateman, L. (Ed.), London, MacLaren Press, 1963, Chapter 16.

41. ANDERSON, D. E. and KOVACIC, P., *Ind. Eng. Chem.*, **47** (1955), p. 171.

42. MAYNARD, J. T. and JOHNSON, D. R., *Rubber Chem. Technol.*, **36** (1963), p. 963.
43. BALDWIN, F. P., BUCKLEY, D. J., KUNTZ, I. and ROBINSON, S. B., *Rubber Plastics Age*, **42** (1961), p. 500.
44. BROWN, H. P. and DUKE, N. G., *Rubber World*, **130** (1957), p. 784.
45. KOVACIC, P., *Ind. Eng. Chem.*, **47** (1955), p. 1090.
46. STIVERS, D. A., in *Rubber Technology*, Morton, M. (Ed.), New York, Van Nostrand Reinhold, 1973, p. 407.
47. HOFFMAN, H. M. R., *Angew. Chem., Int. Ed. Eng.*, **8** (1969), p. 556.
48. BAKER, C. S. L., BARNARD, D. and PORTER, M., *Rubber Chem. Technol.*, **43** (1970), p. 501.
49. BALDWIN, F. P., BORZEL, P., COHEN, C. A., MAKOWSKI, H. S. and VAN CASTLE, J. F., *Rubber Chem. Technol.*, **43** (1970), p. 522.
50. KOVACIC, P. and HEIN, P. W., *Rubber Chem. Technol.*, **35** (1962), p. 528.
51. BATEMAN, L., CUNNEEN, J. I., MOORE, C. G., MULLINS, L. and THOMAS, A. G., in *The Chemistry and Physics of Rubberlike Substances*, Bateman, L. (Ed.), London, MacLaren Press, 1963, Chapter 19.
52. COX, W. L. and PARKS, C. R., *Rubber Chem. Technol.*, **39** (1966), p. 785.
53. TOBOLSKY, A. V. and LYONS, P. F., *J. Polym. Sci., Part A-2*, **6** (1968), p. 1501.
54. LAL, J., *Rubber Chem. Technol.*, **43** (1970), p. 664.